Lecture Notes in Computer Science 9973

Commenced Publication in 1973
Founding and Former Series Editors:
Gerhard Goos, Juris Hartmanis, and Jan van Leeuwen

More information about this series at http://www.springer.com/series/7407

Andrej Brodnik · Françoise Tort (Eds.)

Informatics in Schools

Improvement of Informatics Knowledge and Perception

9th International Conference on Informatics in Schools:
Situation, Evolution, and Perspectives, ISSEP 2016
Münster, Germany, October 13–15, 2016
Proceedings

 Springer

Editors
Andrej Brodnik
University of Ljubljana
Ljubljana
Slovenia

Françoise Tort
ENS Paris-Saclay
Cachan
France

ISSN 0302-9743   ISSN 1611-3349 (electronic)
Lecture Notes in Computer Science
ISBN 978-3-319-46746-7   ISBN 978-3-319-46747-4 (eBook)
DOI 10.1007/978-3-319-46747-4

Library of Congress Control Number: 2016952522

LNCS Sublibrary: SL1 – Theoretical Computer Science and General Issues

Printed on acid-free paper

This Springer imprint is published by Springer Nature
The registered company is Springer International Publishing AG
The registered company address is: Gewerbestrasse 11, 6330 Cham, Switzerland

Preface

This volume contains the papers presented at the 9th International Conference on Informatics in Schools: Situation, Evolution and Perspective – ISSEP 2016. The conference, held during October 13–15, was hosted at the University of Münster, Germany. The ISSEP series started in 2005 in Klagenfurt. It was followed by meetings in Vilnius (2006), Torun (2008), Zürich (2010), Bratislava (2011), Oldenburg (2013), Istanbul (2014), and Ljubljana (2015).

The conference focuses on educational goals and objectives of informatics or computer science as a subject matter in primary and secondary schools (K-12 education) and their different realization in compulsory and voluntary courses. It provides an opportunity for researchers and educators to reflect upon the goals and objectives of the subject, its curricula and various teaching and learning paradigms and topics, possible connection to every day life, and various ways of establishing informatics education in schools. Consequently, the papers published in this volume present different aspects of computer science education and in particular computer science teaching with a goal to *improve informatics knowledge*, and what is particular interesting, *to change perception and attitude towards informatics and/or computer science*. Papers address many educational topics including teaching and learning materials, teacher training, various forms of assessment, traditional and innovative educational research design, motivating competitions like Bebras, and we are very happy that they also touch issues such as the motivation of girls for computer science.

This year, the conference was held together with the 11th Workshop in Primary and Secondary Computing Education – WiPSCE 2016. It gave the opportunity to bring together both communities and get a broader dissemination of the results.

The conference received 50 submissions. Each submission was reviewed by at least three Program Committee members and evaluated with respect to its quality, originality, and relevance to the conference. The committee decided to accept 17 papers to be published in the LNCS proceedings, which corresponds to 34 % of received papers. The decision process was made electronically using the EasyChair management system. And last but not least, since the ISSEP was colocated with the WiPSCE, this, besides bringing together a larger community and giving both conferences a bigger dissemination impact, also made it possible to have three invited talks by Marc J. de Vries, by Raymond Lister, and by Gilles Dowek. The abstract of the last one is also included in this volume.

We would like to thank all those who contributed to this conference becoming a success: the authors who responded to the call for papers, the members of the Program Committee and the additional reviewers who carefully read the papers and wrote reports that allowed authors to improve their submissions, the invited speakers who shared their experience and thinking with the audience. Special thanks also goes to Georges-Louis Baron, who organized the poster session, and Peter Micheuz for the organization of work-shops. We would like to warmly thank Jan Vahrenhold, chair of

WIPSCE, who accepted an extra workload by hosting ISSEP and who managed to make the co-hosting of both conferences a real success. We also thank the members of the Organizing Committee from the University of Münster.

August 2016   Andrej Brodnik
                                                                   Françoise Tort

Organization

Program Committee

Marc Berges	Technische Universität München, Germany
Miles Berry	University of Roehampton, UK
Javier Bilbao	University of the Basque Country, Spain
Andreas Bollin	University of Klagenfurt, Austria
Andrej Brodnik (Chair)	University of Ljubljana and University of Primorska, Slovenia
Valentina Dagiene	Vilnius University, Lithuania
G. Barbara Demo	Università Torino, Italy
Ira Diethelm	Carl von Ossietzky Universität Oldenburg, Germany
Béatrice Drot-Delange	Université Blaise Pascal - Clermont-Ferrand II, France
Michail Giannakos	Norwegian University of Science and Technology, Norway
Yasemin Gulbahar	Ankara University, Turkey
Juraj Hromkovic	ETH Zurich, Switzerland
Ivan Kalas	UCL Institute of Education, UK
Peter Micheuz	Alpen-Adria-Universität Klagenfurt, Austria
Christophe Reffay	University of Franche-Comté, France
Ralf Romeike	Friedrich-Alexander-Universität Erlangen-Nürnberg, Germany
Carsten Schulte	Freie Universität Berlin, Germany
Maciej Syslo	Nicolaus Copernicus University in Toruń, University of Wroclaw, Poland
Françoise Tort (Chair)	ENS Paris-Saclay, France
Mary Webb	King's College London, UK

Posters

Georges-Louis Baron	Université Paris V René Descartes, France

Workshops

Peter Micheuz	Alpen-Adria-Universität Klagenfurt, Austria

Additional Reviewers

Michael Brinkmeier	Claudia Hildebrandt	Jens Maue
Markus Dahinden	Ludmila Jašková	Malika More
Spyros Doukakis	Mehdi Khaneboubi	Stéphanie Netto
Varvara Garneli	Dennis Komm	Sofia Papavlasopoulou
Elio Giovannetti	Pascal Lafourcade	Michal Winczer

Local Organization

Holger Danielsiek	Westfälische Wilhelms-Universität Münster, Germany
Dana Glasmeyer	Westfälische Wilhelms-Universität Münster, Germany
Jan Vahrenhold	Westfälische Wilhelms-Universität Münster, Germany
Mirko Westermeier	Westfälische Wilhelms-Universität Münster, Germany

Sponsoring Institution

École Normale Supérieur	Université Paris-Saclay, France
Westfälische Wilhelms	Universität Münster, Germany

Elements to Define a Coherent Curriculum for the K12 Education: The Example of France (Invited Paper)

Gilles Dowek

Inria, ENS-Cachan, and Société Informatique de France
gilles.dowek@ens-cachan.fr

Since the beginning of this academic year, informatics has been taught at all levels in the French K12 education system. This required, not only to define a curriculum for each level, but also to build a global view of this teaching and of its organization over time. The Scientific Committee of the *Société Informatique de France* has conducted such a reflection.

This talk presents the new situation of informatics in the French education system and some of the conclusions of this reflection.

The main idea is that teaching informatics requires to take into account three forms of complexity of informatics itself. First, informatics is both a science and a technology. Then, it articulates four concepts that existed before it, but that have been completely renewed. Finally, it is the yeast of a dramatic transformation of the world. This requires, when teaching informatics, to take care of three equilibria: between scientific and technological activities, between the concepts, and between the core of the subject and its interfaces.

A Science and a Technology. Informatics is at the same time a science, that allows to know, for instance, that there are no linear time sorting algorithms, and a technology that allows to build, for instance, a program to sort data. The objects built in informatics are often immaterial and their construction requires different skills than in other technologies.

Learning how to write programs is a key step when learning informatics, as, this way, the students become autonomous, and stop using objects built by others, to start building their own. The students can start programming very early, using graphic languages, even before recognizing the letters. But the right time to master programming seems to be middle school.

This allows to divide the K12 curriculum in three major steps: discovering the concepts of informatics in kindergarten and elementary school, mostly using unplugged activities, acquiring programming skills in middle school, learning informatics as a science in high school.

The fact that informatics is both a science and a technology also impacts its pedagogy, that must be project oriented at all levels. Like when learning to play music, practicing is essential when learning informatics.

Four Concepts. A computer is a machine that executes algorithms. Although the concepts of machine and algorithm existed before, they have been completely renewed by informatics. Letting machines execute algorithms also requires to express them in a formal language and to express the objects processed by these algorithms as sequences of symbols. This brought two other concepts to informatics: language and information. These two concepts too existed before informatics, but have been completely renewed. At all levels, these concepts must be taught in a balanced way.

For each of these concepts, we can define a progression over time. For instance, we suggest the following progression for the concept of language. In primary school, the students may discover the notion of language, through a language describing simple dance patterns, like "N3; E4; S3" for: three steps north, four steps east, three steps south, or through the first elements of a programming language. They can also create such languages by themselves. In middle school, the curriculum is focused on programming and programming languages. In high school, they can learn advanced features of programming languages, discover the notion of grammar, and invent and implement their own tiny programming languages.

Similar progressions can be defined for other concepts. The details of the progressions are not important: it does not really matter whether this or that is taught in eighth or in ninth grade. What is important is that they exist, so that the curricula for each level can be defined in a coherent way.

The Yeast of a Transformation of the World. Informatics is the yeast of a dramatic transformation of the world and this transformation is a wonderful lever to motivate the students to learn informatics, and science and technology in general.

Informatics transforms the way the students communicate with their friends and, as this affects them directly, it must be addressed in class. A simplistic, but wrong, solution is to give to the students a list of "dos and don'ts using social media", they would understand neither the origin nor the meaning of. A better approach is to focus on the properties of digital information—easy duplication, quick communication, persistence over time…—and let the students define their own good practices on social media, taking these properties into account.

Some of the questions related to the transformation of society, for instance the transformation of encyclopedias, impact and motivate everyone. Others, for instance the evolution of music composition, motivate only the students already interested in some subject, for instance music. The choice of the topics to be developed thus must be guided by the area of interest of the students.

In one case and in the other, these topics are wonderful opportunities for interdisciplinary projects. The rôle of the informatics teacher in these projects is to relate the scientific and technological knowledge to their impact on society. For instance it is pointless to note that the way photographers work has evolved. But it is fruitful to remark how the digital representation of images has impacted the work of photographers.

Contents

Work in Progress

Research Papers

Teaching Computer Image Processing Subject to Middle School Students: Cognitive and Affective Aspects

Khaled Asad[1,2(✉)]

[1] Alqasemi Academic College of Education, Baqa-El-Gharbia, Israel
kasad@qsm.ac.il
[2] Beit-Berl Academic College of Education, Kfar-Saba, Israel

Abstract. Today's youth are making extensive use of technological devices such as smart phones and computers. These devices are based on inter-disciplinary knowledge. Are these young students attracted to learn the computer principles that these devices are based on? Many educators agree that one of the methods to foster learning in school is to connect the topics of study with students' interests, experiences and daily life, 'contextual learning'. This paper describes a research aimed at examining the case of teaching a course on computer image processing to middle school students, and evaluating its influence on students cognitively and effectively. The study included the development, implementation and evaluation of a computer image-processing course. The course was taught to 34 9th-grade students in two groups. The control population comprised 64 9th-grade students in three groups. The study included developing an instructional model consisting of four phases: teaching theory, manual and computerized practices, implementing challenging tasks, and projects. Data were collected by using quantitative and qualitative research tools, such as two exams, three projects and a half-opened attitude questionnaire about learning computers, class observations and semi-structured interviews with students and teachers. Findings showed that young students' achievements were very well in learning principles of image processing. In the mathematics exam, the experimental students' achievements were significantly higher than the control students' achievements. The students showed high motivation and great interest in learning the course. Finding showed that the instructional model developed in the study was the main component influencing the experimental students' achievements and motivation.

Keywords: Computer science education · Contextual learning · Interdisciplinary learning · Constructive learning · Mathematics in context

1 Introduction

The educational literature strongly supports the notion of 'contextual learning', which is about engaging students in learning subjects that interest them and close to their world and daily lives. Mathematics, science and technology are taught in school as separated subjects and students do not see the connection between them. Furthermore,

© Springer International Publishing AG 2016
A. Brodnik and F. Tort (Eds.): ISSEP 2016, LNCS 9973, pp. 3–13, 2016.
DOI: 10.1007/978-3-319-46747-4_1

today's youth are making extensive use of advanced technological devices such as cellphones, digital cameras and computers. Teaching basic concepts on which these modern devices are based on could serve as a good platform for fostering student's interest in technology and developing their higher-order intellectual skills such as problem-solving and creativity.

This paper describes a research study that included the development, implementation and evaluation of a scientific-technological course on computer image processing, which is considered a very challenging field [9]. The research was guided by the following questions: To what extent can middle school students with no background in computers learn an advanced scientific-technological subject such as image processing? How would such a course affect their perceptions about the subject matter, their motivation and interest to pursue a career in computers and technology?

2 Theoretical Background and Related Work

This section first addresses some issues from the educational literature relating to teaching advanced scientific-technological subjects to young students. After then, it reviews the contents of the image processing course that was developed and explored in the research.

2.1 Contextual Learning

The term contextual learning is mainly about learning that relates to a learner's diverse life contexts such as at home, leisure time, social or environmental activities, or the work place [7]. Contextual learning is not only about what students learn but also how they learn. They learn best when they deal with subjects that related to their own lives and interests [4]. To gain the best of contextual learning and to achieve significant learning, the problem or the task should be driven by a question that opens a door to make a connection between activities and the related underlying conceptual knowledge [8]. See Sect. 2.3 bellow.

2.2 Interdisciplinary Learning

Interdisciplinary learning is about providing the students with opportunities and space for learning beyond subject boundaries and making connections between different areas of learning [15]. Educators in the field of science and technology increasingly recognize the need to develop curricula that combine learning issues in science, technology, engineering and mathematics (STEM - Science, Technology, Engineering and Mathematics). Recently, this approach is intended to reflect the nature of science and technology, and to increase students' interest in learning these subjects [6]. Advocates of more integrated approaches to K–12 STEM education argue that teaching STEM in a more connected manner, especially in the context of real-world issues, can make the STEM subjects more relevant to students and teachers [12].

2.3   Knowledge Types

Educators emphasize the importance of recognizing the types of knowledge in order to assess the knowledge acquired by students [11]. In general, it is accepted to distinguish between three types of knowledge: factual knowledge, procedural knowledge and conceptual knowledge [1, 11], as follows:

Factual Knowledge is the part of knowledge that describes information such as names of people, places, dates and events. It is about to know "what". For example, many people "know" that their phone digital camera has "8MP", but maybe, they do not understand the real meaning of this figure.

Procedural Knowledge clarifies how to do things according to the rules, laws, formulas or algorithms. It is about to know "how to do". For example, how to find the roots of a quadratic equation, or how to calculate the equivalent resistance of resistors connected in parallel circuit.

Conceptual Knowledge is the knowledge of the relationships and interactions between knowledge items. It is more complex and more organized than factual knowledge, and reflects a deep understanding of content. It is about to know "why". For example, understanding the concept of "energy" in physics, chemistry and biology; or understanding the concepts of "ratio and scale" in mathematics and physics. Conceptual knowledge is acquired by a prolonged study and experience, and cannot be learned directly [3, 11, 14].

The taxonomy of knowledge types presented above represents another dimension of Bloom's taxonomy but is not intended to replace it [1].

The image processing course examined in this research was designed to put into practice some of the ideas reviewed above, contextualizing learning in subjects that are personally meaningful to the students, integrating the learning of subjects in science, technology and mathematics, and creating a constructivist learning environment in which students deal with challenging tasks and having opportunities for peer learning. However, one must be aware that introducing advanced technological subjects such as image processing into the school curriculum is not a simple task due to the complexity of the subject and the need for integrating knowledge from a number of disciplines. Therefore, the course was designed and delivered according to an instructional model that combine short instruction periods by the teacher and task-based learning by the students, as will be detailed later in this paper. Consequently, this research aimed at exploring students' learning, achievements and attitudes towards learning the new subject. More specifically, the research was guided by the following questions:

- What is the impact of learning a course in computer image processing on students in terms of their achievements in learning the subject and in mathematical aspects regarding factual, procedural and conceptual knowledge?
- What is the effect of learning the course on students' motivation to learn computers at school, and their interest to pursue a computer career in the future?
- What elements of the curriculum and teaching method contributed to or subtracted from learning the course?

3   The Computer Image Processing Course

In the presented study, we developed a course on computer image processing that included the following topics:

- Digital representation of an image: binary and decimal numbers, pixels, resolution, colors.
- Image processing and enhancement, such as: producing a negative image, amending and changing the image brightness and contrast by using simple and complex mathematical operations, such as adding, multiplication and doing histogram equalization.
- Image formats such as bmp, gif, jpg, png and compression methods such as RLE.
- Advanced mathematical operations for image processing such as, spatial filtering to remove noise according to the average or median and creating artistic effects.
- Facial recognition – how a computer program can identify an individual by comparing its picture to other pictures stored in a database.

3.1   Computer Image Processing as an Interdisciplinary Subject

Computer image processing subject combines knowledge from the fields of science, mathematics and computer science [13, 17, 18]. Following a primary example from the course shows the strong connection between image processing and mathematics.

Example: Black-White and Color Images as Matrices of Numbers. An image we see on a computer screen is a collection of pixels that are stored in a computer memory as a matrix of numbers. Figure 1 illustrates how a small part of a picture is represented as a matrix of n × m numbers. Each number in the matrix represents a pixel brightness level. In black-white images, each pixel is represented by a value that ranges between (0–255), where 0 represents (full black) and 255 (full white).

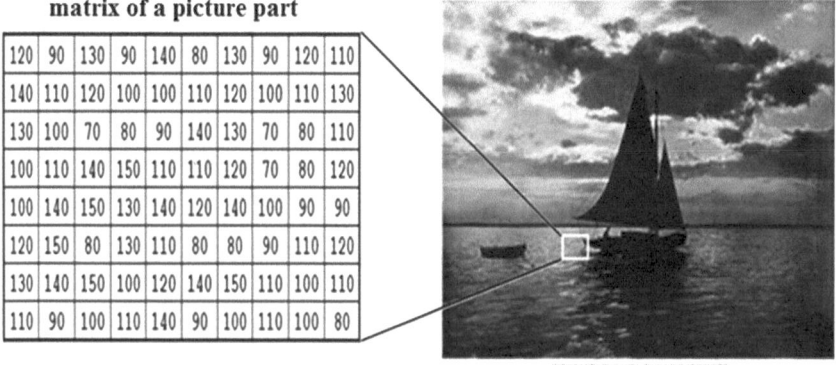

Fig. 1. Picture is composed of tiny pixels and each pixel has its own value that represents a brightness level

The numbers 0, 100 and 150 in Fig. 1 are decimals. Students learn the smallest unit of information in a computer is a bit that is represented by the values 0 or 1. The brightness of each pixel is represented by 8 bits = 1 byte. Students learn how to perform conversions in binary to decimal and vice versa. For color images, each pixel is represented by three numbers in the field (0–255) which represent the three primary colors: red (R), green (G) and blue (B). Therefore, a color image is represented by three matrices as the one above. Computer image processing operations, such as changing the brightness and contrast, moving, mirroring, rotating and noise removing are all implemented by mathematical operations on the values of the pixels.

The example presented above illustrates the deep connection between the subject of image processing and mathematics. Hence, one of the goals of this study was to examine the effect of teaching the course on student achievements in mathematical concepts related to the topics learnt in the course.

4 The Study

4.1 The Study Plan and Objectives

The central axis of the study involved the development, implementation and evaluation of a course on computer image processing principles for middle school students. The study is aimed to evaluate student achievements in learning the principles of computer image processing and mathematical aspects related to the subject, with respect to three types of knowledge: factual, procedural and conceptual. The research also examined the impact of the course on students' attitudes in terms of interest and motivation in learning the subject.

4.2 The Study Population

The study took place in a middle school located in city in northern Israel. The study population comprised of two experimental groups (n1 = 34) and three control groups (n2 = 64). All students in the groups are of 9th grade. It's important to say that all 9th grade classes in school are similar and heterogeneous in terms of students' achievements level in each class, according to sorting tests. This allowed us to have similar groups of experimental and control.

4.3 Teaching the Course

Two qualified computer teachers have taught the course to the two experimental groups under the supervision of the researcher. The course lasted 15 sessions of 90 min each and included the study of theoretical and practical learning based on doing exercises, challenging tasks and projects. As a result of the findings of the pilot study [2] conducted earlier under similar conditions, in the current study we decided to reduce the theoretical learning part and to increase the learning based on doing tasks and projects. In the light of the experience gained in the mentioned pilot study, we developed an instructional model, as shown in Fig. 2.

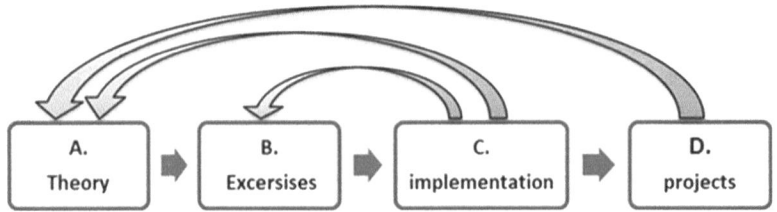

Fig. 2. Instructional model for teaching scientific-technological subjects

The developed instructional model (TEIP) consists of four phases: theoretical instruction and demonstrations (*Theory*); Manual and computerized practice (*Exercises*) at a basic level; Performing challenging tasks and an advanced application by professional software (*Implementation*); and Projects (*Projects*).

According to this model, the teaching-learning goes like this: For each topic, the teacher gives short explanation and demonstrates a new topic with a presentation, for 25–30 min (Theory). Then, for 30 min, the students practice the theoretical learned topic manually by hand-worksheet, and do exercise on computer (Exercises). 30 min left, the students perform challenging tasks and apply the topics learned on a computer by professional software (implementation). For every three to four sessions students receive a comprehensive with larger scale tasks as a (project).

In the study and during the course, the experimental students learned authentic topics, performed preliminary exercises and submitted three projects on topics from real-world examples and related to students' daily lives, such as image enhancement, photographing and measuring building height, and facial recognition. The control students have not studied the image-processing course; however, they shared the experimental students the same mathematics regular classes.

4.4 Methodology and Data Collection Tools

In order to assess the impact of students' learning course cognitively and effectively, the study combined quantitative and qualitative methods aimed at collecting as much information as possible on students' activities in the class, their achievements and their attitudes towards the course. As a qualitative tools we used a half close-ended questionnaire and two achievement exams: one exam in image processing principles for the experimental group and the second exam in the related mathematical topics targeted both the experimental and the control groups; as a quantitative tools we used class observations, interviews with students and teachers, and analyzing three projects.

5 Findings

5.1 Achievements in Learning Image Processing Principles

The students' achievements in image processing were evaluated by a 90 min comprehensive exam that conducted at the end of the course and through the analysis and evaluation of three projects that were submitted by the students. The exam was composed by the researcher and has been validated by three experts in computer science education. They examined the validity of the exam in two aspects: examining the exam questions according to the material content of the course; and the knowledge types that the exam questions should test. The exam is comprised of seven questions that test three types of knowledge: factual (38 %), procedural (32 %) and conceptual (30 %).

All the experimental students participated in the image processing exam (N = 34). The average score of students in three types of knowledge (scale 0–100), were as follows: factual knowledge $\bar{x} = 78.2$ (SD = 14.95); procedural knowledge $\bar{x} = 83.1$ (SD = 21.43); conceptual knowledge $\bar{x} = 69.5$ (SD = 21.52).

From these results, we learn that middle school students have learned and dealt well with the study of the subject, although their achievements were lower in the parts that examined conceptual knowledge than the achievements in the parts that examined procedural and factual knowledge.

5.2 Achievements in Project Work

As mentioned above, during the course the students performed three projects involving semi-open tasks, as follows: The first project was about measuring the height of objects using a digital camera. The second project was about shooting and enhancing pictures. The third project was about face recognition. For example, the students in the second project took pictures of things or scenes in different lighting conditions and improved the pictures by mathematical operations. The students were given a full explanation and details about the requirements and the learning objectives of each project.

The researcher and two teachers evaluated the students' projects submitted according to three assessment indicators. Each indicator covers some aspect to be considered in each project according to its educational goals. Evaluating the projects show that the students' achievements on a scale (0–100) were as follows: Measuring the height of objects $\bar{x} = 87.6$ (n = 33, SD = 12.09); Enhancement of pictures $\bar{x} = 80.9$ (n = 32, SD = 16.23); Face recognition $\bar{x} = 75.0$ (n = 29, SD = 8.27).

From these overall students' achievements in projects, we learn that the students have shown a good ability to deal with complex tasks on the subject. The first project students' achievements were better than their achievements in the second and third project, which were more complex.

5.3 Achievements in Learning Mathematical Concepts

The students' achievements in mathematics were evaluated by a 75-min exam that was conducted at the end of the course for both the experimental and control students. The exam was validated by three experts in mathematics education. They examined the validity of the exam in two aspects: examining the exam questions to match the mathematics curriculum in middle school, and examining the types of knowledge that the exam questions should test. The exam included four questions that examine two types of knowledge: procedural (56 %) and conceptual (44 %). The exam was given in parallel for both, the experimental and control groups in the school. In the exam were participated 33 students from the experimental group and 64 students from the control. To ensure the evaluation reliability of the exam, it was evaluated by the researcher and two teachers of mathematics according to an assessment indicator. The graph in Fig. 3 shows the mean scores of the students in the mathematics exam.

By analyzing the school average scores in mathematics of the study's population, we found no significant differences between the control and experimental groups. This finding indicates that the experimental and control groups have the same background in mathematics.

The scores of the students in the mathematics exam developed in the study show that the scores of the experimental group were significantly better than the scores of the control group in procedural knowledge ($t_{(95)} = 2.26, p < .05$), in conceptual knowledge ($t_{(95)} = 5.95, p < .05$) and in the average total scores of the exam ($t_{(95)} = 4.47, p < .05$). These findings indicate a positive impact of learning computer image processing topic on student achievements in mathematics.

The graph in Fig. 3 shows that while the gap between the conceptual and the procedural knowledge average score is about 21 points in the experimental group, this gap is approximate 37 points in the control groups. This finding could indicate that the experimental students have acquired conceptual knowledge of the mathematical topics that were tested in the exam are beyond the ordinary materials that have been taught in math classes at school. That is, due to the research course, the significant change in the experimental students compared to students in the control groups was gaining conceptual knowledge. This is related to the fact that the mathematical concepts in the course were learnt in context of image processing.

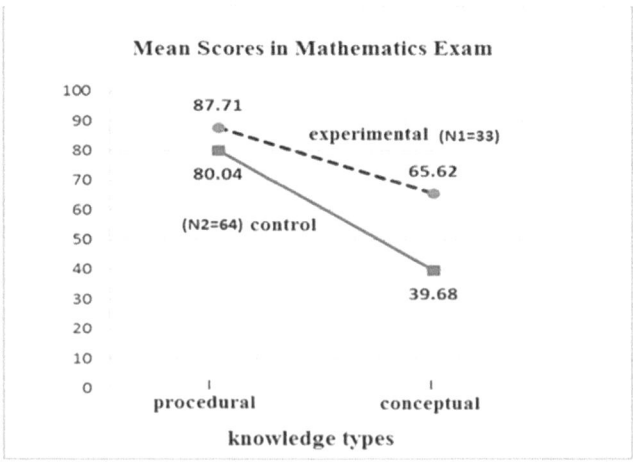

Fig. 3. The average scores for the experimental and control groups in mathematics exam

5.4 Students' Attitudes

In addition to the achievement results mentioned above, the study examined the attitudes of students in terms of their motivation to study the subject matter, to study computers at school, and their wellness to work in the field in the future. The study examined these aspects by qualitative instruments such as, class observations and interviews with students, and by quantitative tools such as, semi-open ended questionnaire about learning computer. For convenience and briefness matters in this article, we present few of the findings on these issues.

Students' Attitudes During Learning Theory Stage and Project-Based Learning Stage. At the stage of theoretical study, the findings from classroom observations and the interviews with students showed that students exhibited medium to high level of interest and motivation to learn. However, as long as the theoretical topic being studied was close to the world of students and dealt with issues related to their daily lives, their interest and motivation to learn were higher. Throughout the study, results showed that students preferred the practical learning than the theoretical learning. In particular, they enjoyed learning when they performed exercises on a computer with professional software, got engaged with challenging tasks and prepared individual projects that were meaningful to them. Throughout these activities, the students demonstrated greater interest and motivation in learning the course.

Here is an example of an interview with a student: Interviewer: *"Tell me, what did you like in the course?"* Student: *"the activities"*. Interviewer: *"What activities? Give an example."* Student: *"Coloring a picture that was in black and white.. I never thought that's possible to add colors and color black and white pictures."* Interviewer: *"What else interested you?"* Student: *"..that all natural colors can be obtained from only three colors* [Red, Blue, Green], *at first I did not believe it was real, till I saw and tried it on computer. This encouraged me to learn more."* This example represents what many students have said and wrote about the course in the open section of the attitude questionnaire.

In their reflection on the projects, some students wrote: *"Working on the project contributed to my knowledge and motivated me to do things on my own at home"*; *"At first, we thought this subject was difficult, now after completing the project we saw how easy it is"*; *"I would like to work in this profession and learn more about computers and how it works."*

In addition to the above, the teachers who taught the course pointed out that the students got interested most when they got involved with challenging tasks and projects. Another interesting finding was that while learning the image processing course, the students naturally utilized mathematical concepts that were new for them without any special difficulties.

6 Discussion and Conclusions

The findings showed that middle school students were able to study the principles of computer image processing, which is rich scientific-technological and interdisciplinary subject, after adjusting the course contents to their previous knowledge in science and mathematics. The two main factors that contributed well to the success and the motivation of the students are:

- The instructional model (*Theory, Exercise, Implementation,* and *Project* - TEIP) that was developed in the study that reducing the weight of teacher formal instruction or solving pre-designed assignments and increasing the students engagement in challenging tasks and open-ended assignments and projects.
- Allowing students to be engage in authentic content and to carrying out practical tasks related to their world, such as improving their own images or applying the facial recognition task to pictures of their classmates.

These findings are compatible with the theory of contextual learning in which students acquire knowledge and learn well as they study topics that meaningful for them and related to the real world [5]. The research findings also indicate the contribution of Project-based learning (PBL) to help students make the connection between what they learn in school and the real world outside [10, 16, 19]. Student success in learning mathematical knowledge related to computer image processing topic, highlights the contribution of interdisciplinary learning that combine studies in science, technology, engineering and mathematics (STEM) [6].

In conclusion, the study findings highlight that, despite the fact that computer image processing is complex, in terms of students' learning it was helpful, interesting and challenging. This suggests that there is room to integrate teaching scientific-technological subjects such as image processing, robotics or computer science in middle school. However, it should be guided by constructive pedagogy while reducing the weight of formal teacher instruction or engaging the students in solving pre-designed exercises.

References

1. Anderson, L.W., Krathwohl, D.R. (eds.): A Taxonomy for Learning, Teaching and Assessing: A Revision of Bloom's Taxonomy of Educational Objectives: Complete Edition. Longman, New York (2001)
2. Asad, K., Barak, M.: Teaching image processing concepts in junior high school: the role of student-centered vs. traditional instruction. In: Technological Learning and Thinking conference (TL&T), University of British Columbia, 17–21 June 2010
3. Ben-Hur, M.: Concept-rich Mathematics Instruction: Building A Strong Foundation for Reasoning and Problem Solving. Association for Supervision and Curriculum Development (ASCD), Alexandria (2006)
4. Brandt, R.S.: Powerful Teaching and Learning. Association for Supervision and Curriculum Development, Alexandria (1998)
5. Brown, J.S., Collins, A., Duguid, P.: Situated cognition and the culture of learning. Educ. Res. **18**(1), 32–42 (1989)

6. Bybee, R.W.: Advancing STEM education: a 2020 vision. Technol. Eng. Teach. **70**(1), 30–35 (2010)
7. Dewey, J.: Experience and Education. The Kappa Delta Pi Lecture Series. Macmillan Publishing Company, New York (1963)
8. Dolmans, D.H., De Grave, W., Wolfhagem, I.H., Van Der Vleuten, C.P.: Problem based learning: future challenges for educational practice and research. Med. Educ. **39**(7), 732–741 (2005)
9. Gonzales, R.C., Woods, R.E.: Digital Image Processing. Prentice Hall, Upper Saddle River (2002)
10. Hmelo-Silver, C.E.: Problem-based learning: what and how do students learn? Educ. Psychol. Rev. **16**(3), 235–266 (2004)
11. McCormick, R.: Issues of learning and knowledge in technology education. Int. J. Technol. Des. Educ. **14**(1), 21–44 (2004)
12. National Academy of Science (NAS): STEM Integration in K-12 Education, Status, Prospects, and An Agenda for Research. National Academies Press, Washington, DC (2014). http://www.nap.edu/catalog.php?record_id=18612
13. Oldknow, A.: Mathematics from still and video images. Micromath **19**, 30–34 (2003)
14. Rittle-Johnson, B., Alibali, M.W.: Conceptual and procedural knowledge of mathematics: does one lead to the other? J. Educ. Psychol. **91**(1), 175–189 (1999)
15. Rowntree, D.: A Dictionary of Education. Barnes and Noble, Totowa (1982)
16. Savery, J.R.: Overview of problem based learning: definitions and distinctions. Interdisc. J. Probl. Based Learn. **1**(1), 9–20 (2006)
17. Silverman, J., Rosen, G.: Supporting students interest in mathematics through applications from digital image processing. J. Res. Center Educ. Technol. **6**, 63–77 (2010). http://www.rcetj.org/index.php/rcetj/article/view/138
18. Tanimoto, S., King, J., Rice, R.: Learning mathematics through image processing: constructing cylindrical anamorphoses. In: Proceedings of MSET 2000, International Conference on Mathematics/Science Education and Technology, San Diego, CA, 5–8 February 2000
19. Thomas, J.W.: A Review of Research on Project Based Learning. Autodesk, San Rafael (2000). http://www.bie.org/files/researchreviewPBL.pdf

Analyzing Conceptual Content of International Informatics Curricula for Secondary Education

Erik Barendsen[1]([envelope]) and Tim Steenvoorden[2]

[1] Radboud University and Open University, Nijmegen, The Netherlands
e.barendsen@cs.ru.nl
[2] Radboud University, Nijmegen, The Netherlands
t.steenvoorden@cs.ru.nl

Abstract. Various countries are in the process of curriculum innovation with respect to informatics, which makes it interesting to conduct a systematic international comparison. As a first step, we focus on the analysis of conceptual content of curriculum specifications, that is, formal descriptions and guidelines. As a case study, we apply our method to analyze five curriculum specifications, including the former (2007) and new (2016) Dutch informatics curriculum for upper secondary education. The results indicate interesting similarities and differences with respect to emphasis of specific conceptual areas such as algorithms, software engineering and social aspects. The method appears fruitful to determine, for example, the position of the new Dutch curriculum relative to the former curriculum and to the three other recent international specifications.

Keywords: Curriculum · Concepts · Content analysis

1 Introduction

In the past few years, several organizations and individuals in Europe and the United States have expressed concerns about the state of informatics education (Académie des Sciences 2013; Furber 2012; Gander et al. 2013; Kaczmarczyk and Dopplick 2014; KNAW 2012; Samaey et al. 2014).

Although the underlying motivations vary, the common outcome of the above reports is that our society is becoming more and more digitized and therefore a broad group of people (especially children) need to learn about ICT as well as the skillful and responsible use of digital tools. Moreover, interested young people should get the opportunity to receive further education in informatics.

Various countries are in the process of curriculum innovation or have recently completed such a reform. The developments have been documented in formal curriculum documents and in guidelines.

In *England*, for example, a new subject Computing has been introduced for all students (British Department for Education 2013). The organization Computing at School developed guidelines for the new subject (Computing at School Working Group 2012). The *US* teacher organization CSTA published standards

© Springer International Publishing AG 2016
A. Brodnik and F. Tort (Eds.): ISSEP 2016, LNCS 9973, pp. 14–27, 2016.
DOI: 10.1007/978-3-319-46747-4_2

for K–12 computer science (CSTA 2011). In *France*, an informatics curriculum has been introduced for grades 9–12 (Ministère de l'Éducation Nationale 2012). The current informatics curriculum in *The Netherlands* for grades 10–12 dates from 1998 and has not changed since then, except for a minor reformulation in 2007 (Grgurina and Tolboom 2008). In March 2016, a new curriculum proposal, commissioned by the Ministry of Education, was completed (Barendsen and Tolboom 2016).

The variety of developments make it interesting to conduct a systematic content analysis of curricula and guidelines, in order to support international comparison and curriculum development. However, it is not easy to compare the above curriculum documents, as the composition, length, and formulation of the specifications vary a lot.

In order to support the development of a new curriculum in the Netherlands, we were interested in the conceptual content (i.e., topics and ideas belonging to the informatics subject matter) of existing curricula and guidelines.

The so-called *Darmstadt Model* is a more general framework for classifying implementations of informatics education in various countries (Hubwieser et al. 2011; Hubwieser 2013). Our analysis relates to the categories *knowledge* and *intentions* within the dimension *educational relevant areas* of the Darmstadt Model. In the process of developing the framework, Hubwieser et al. (2011) perform a global categorization of the learning objectives in four countries using the ACM classification scheme and the CSTA strands as categories.

In our study, we aimed for a detailed and in-depth analysis of concepts, regardless of the skills or attitudes in which they appear (cf. Barendsen et al. 2015).

An alternative type of conceptual content analysis, based on a survey among local experts, was part of a comparison of teaching practices in Germany and the UK (Dagiene et al. 2013).

2   Aim of the Study

Part of the research was carried out during the construction of the new Dutch informatics curriculum, aiming at positioning the ideas of the curriculum committee in international perspective (Steenvoorden 2015). The starting point of the curriculum development was an international workshop in September 2014 at the *Lorentz Center* at Leiden University in the Netherlands. The curricula and documents discussed in the workshop constituted the first sample for our analysis: the former Dutch curriculum (Schmidt 2007), the French informatics curriculum (Ministère de l'Éducation Nationale 2012), the CAS guidelines (Computing at School Working Group 2012), and the CSTA standards (CSTA 2011).

The other part of the research was conducted after completion of the new curriculum (Barendsen and Tolboom 2016), to determine similarities and differences between the new curriculum and the other curricula investigated thus far.

The Dutch informatics subject only spans upper secondary education (grades 10–12). The French curriculum is intended for a similar range (grades 9–12). CAS and CSTA constructed guidelines for grades K–12. For a proper comparison, we decided to analyze the latter documents as a whole instead of their respective 10–12 segments, since it is reasonable to expect that some basic concepts (comparable to those found in the Dutch and French 10–12 curricula) appear in the K–9 part of CAS and CSTA documents.

Our research question was: *How can the conceptual content of the new Dutch curriculum, the former Dutch curriculum, the French curriculum, and the CAS and CSTA guidelines be characterized?*

3   Method

We used a variant of the method developed in Barendsen, Fisser, Krüger, and Tolboom (2014) and Steenvoorden (2015), also applied by Barendsen et al. (2015).

Our starting point was a classification of informatics subjects in terms of knowledge categories, based on the 'knowledge areas' of the Computing Curricula (2013). These knowledge areas were developed for higher education, but can be applied fruitfully in our case, since they are complete, that is, certainly cover the secondary education topics. Moreover, the knowledge area descriptions contain detailed specifications, which adds to the reliability of the analysis. The knowledge areas have been clustered into a conveniently small number of categories while maintaining sufficient detail to distinguish variations in content, see Table 1.

We applied an open coding procedure (Cohen, Manion and Morrison 2013) to the documents to extract literal concepts from the curriculum texts. In a second (more axial, cf. Cohen et al. (2013)) coding phase, similar codes were merged into one, slightly more abstract, code. Then the resulting codes were grouped into the general knowledge categories mentioned earlier.

The authors coded samples of the documents (10%) together, while discussing and documenting the code descriptions. Then the remaining texts were coded by the second author. About half of these were reviewed by the first author. Coding differences were discussed and whenever necessary, the category descriptions were refined to reflect the consensus reached in the discussions.

For the analysis, the resulting codes were first used to get a global overview of occurrences of codes in each category. We regard the distribution of occurrences over the categories as an indication of the relative importance of the categories. Then we conducted a more qualitative, in-depth content analysis with respect to selected categories, using the (relative) frequencies and codes as pointers to relevant text segments.

Table 1. Knowledge categories

Knowledge category	Included ACM/IEEE knowledge areas
Algorithms	Algorithms and complexity (AL)
	Parallel and distributed computing (PD)
	Algorithms and design (SDF/AL)
	Remark: concepts about data structures are covered by *Data*
Architecture	Architecture and organization (AR)
	Operating systems (OS)
	System fundamentals (SF)
Modeling	Computational science (CN)
	Graphics and visualisation (GV)
Data	Information management (IM)
	Fundamental data structures (SDF/IM)
Engineering	Software engineering (SE)
	Development methods (SDF/SE)
	Remarks: also contains ideas on collaboration; concepts without an engineering component are covered by programming
Intelligence	Intelligent systems (IS)
Mathematics	Discrete structures (DS)
Networking	Networking and communication (NC)
Programming	Programming languages (PL)
	Platform based development (PBD)
	Fundamental programming concepts (SDF/PL)
Security	Information assurance and security (IAS)
	Remark: concepts about privacy are covered by society
Society	Social issues and professional practice (SP)
Usability	Human-computer interaction (HCI)

4   Results

We present our results in two ways. Firstly, in Table 2 we list the categories for each curriculum, sorted according to (absolute) number of concept occurrences. Secondly, we show the (relative) distribution of concepts across the categories for every document in Fig. 1. The new Dutch curriculum consists of a core curriculum and a number of elective themes. Below, we distinguish between the core curriculum and the curriculum as a whole (including the elective themes).

The total number of concept occurrences (i.e., coded quotations) is given at the bottom of each list in Table 2. The reason that France and the Netherlands

Table 2. Lists of knowledge categories for each curriculum document, sorted from most to least occurring concepts. The number of concept occurrences in each category is displayed between parentheses. The total number of concept occurrences in the document is given at the end of each list.

CSTA	CAS	France
1. Algorithms (44)	1. Algorithms (44)	1. Data (28)
2. Engineering (40)	2. Networking (40)	2. Programming (15)
3. Architecture (37)	3. Architecture (38)	3. Architecture (14)
4. Society (30)	4. Data (33)	Networking (14)
5. Networking (27)	5. Programming (19)	4. Algorithms (13)
6. Programming (25)	6. Engineering (17)	5. Mathematics (8)
7. Data (23)	7. Mathematics (5)	6. Society (5)
8. Security (13)	8. Security (4)	7. Engineering (4)
9. Modeling (12)	9. Society (2)	Modeling (4)
10. Intelligence (11)	10. Intelligence (1)	8. Intelligence (2)
11. Mathematics (8)	11. Modeling (0)	9. Rest (1)
12. Usability (2)	Rest (0)	10. Security (0)
13. Rest (0)	Usability (0)	Usability (0)
(Total: 272)	(Total: 203)	(Total: 108)

Netherlands 2007	Netherlands 2016 (core)	Netherlands 2016 (complete)
1. Architecture (13)	1. Programming (18)	1. Programming (22)
2. Data (12)	2. Engineering (17)	2. Architecture (19)
3. Engineering (10)	3. Data (11)	Society (19)
4. Networking (4)	4. Society (10)	3. Data (18)
Rest (4)	5. Architecture (9)	Engineering (18)
5. Programming (3)	6. Security (7)	Usability (18)
6. Usability (3)	7. Algorithms (6)	4. Security (16)
7. Modeling (2)	8. Usability (3)	5. Algorithms (14)
8. Security (1)	9. Networking (2)	6. Networking (11)
9. Algorithms (0)	10. Intelligence (0)	7. Modeling (7)
Intelligence (0)	Mathematics (0)	8. Mathematics (4)
Mathematics (0)	Modeling (0)	9. Intelligence (3)
Society (0)	Rest (0)	10. Rest (0)
(Total: 52)	(Total: 83)	(Total: 169)

have less coded concepts, is that the learning goals are formulated in a relatively compact way and concepts often are mentioned only once. The CAS and the CSTA documents formulate their guidelines in a more spiral-like way, first formulating learning goals for lower grades and after that for higher grades.

Figure 1 provides a global overview of the five documents and how they compare on the twelve respective knowledge categories and a rest category.

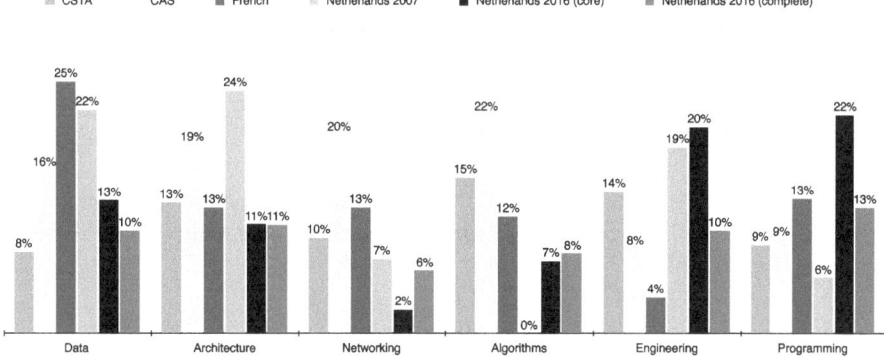

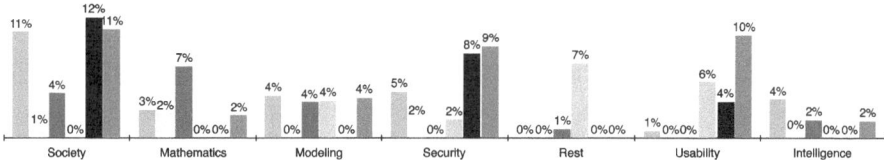

Fig. 1. Relative distribution of concept occurrences across the knowledge categories. The percentages show the fraction of the concept occurrences to the respective categories. For example, 25 % of the concept occurrences in the CSTA guidelines concerns data, while 7 % is about modeling. Categories are sorted by average occurrence.

The frequencies show that *data, architecture, networking, algorithms* and *engineering* cover the biggest parts of the studied specifications.

In this paper we will highlight some interesting differences. Firstly, we note the focus on *data* in the French curriculum and the gap until the next category, *programming*, as we can see in Table 2. Next, the CAS guidelines have the highest score on *algorithms*. Algorithmic concepts appear frequently in several curricula and guidelines. The old Dutch curriculum does not mention any concepts from this category, however. Another interesting observation with respect to the top five categories is the variation in scores within the *engineering* category. For this category, the French curriculum has lower scores than the other documents. Furthermore the high percentages with respect to *society* in the new Dutch curriculum and the CSTA guidelines are remarkable. Finally, the high score of the old Dutch curriculum in the rest category is exceptional.

Below, we will analyze the above observations in more depth. We illustrate our findings with characteristic quotations from the curriculum. In the case of the Dutch and French curricula, we have translated the original texts into English.

4.1   Data

The code frequencies suggest that the French curriculum has the highest emphasis on data (25 %), with *programming* appearing next in the ranking (13 %). This difference of 12 % (13 concepts) may be explained by the structure of the curriculum. Almost all (19 of 28) of the coded concepts in the Data category appear in the section 'Representation of Information'. This is the biggest section in the curriculum description, containing more than a third of the total learning objectives (8 of 21). Of the remaining concepts, 7 appear in the section on 'Languages and Programming'.

In the section on 'Representation of Information', the French curriculum includes objectives about *document formats* and *directory structure*, which furthermore appear only in the CSTA standards.

> *"Formats: Digital data is arranged to facilitate storage and processing. The structuring of digital data respects either de facto standards or norms. Skills: Identify some document formats, images and sound data. Choose an appropriate format compared to a use or need, quality or limitations."* (France)

The curriculum also mentions explicitly that students should learn about the representation of characters, text, numbers, floating points and images.

> *"Digitalization: The computer handles only numeric values. A digitalization step of physical world objects is essential. Skills: Encode a number, a character through a standard code, a text in the form of a list of numeric values. Encode an image or sound as an array of numeric values. [...]"* (France)

The CAS and CSTA curricula only mention *information representation* in general terms.

> *"Analyze the representation and trade-offs among various forms of digital information."* (CSTA, p. 18)

The old Dutch curriculum does not contain any objectives regarding *information representation*. The new Dutch curriculum however, specifies the ability to use *standard representations*.

> *"The candidate is able to use standard representations of numerical data and media, and is able to relate these to each other."* (Netherlands 2016)

In the section on 'Languages and Programming', the French curriculum explicitly states which *data types* students should master.

> *"Data types: Integer; floating point; boolean; natural number; array; string. Skills: Choosing a data type based on a problem to solve."* (France)

In contrast, the new Dutch curriculum refrains from explicitly mentioning specific data types. The same holds for the CSTA guidelines.

"The candidate is able to represent data in a suitable data structure, keeping the purpose in mind; the candidate is able to compare the elegance, efficiency and implementability of various representations." (Netherlands 2016)

When going down to the bit level, the CSTA prescribes the following objective.

"Demonstrate how 0s and 1s can be used to represent information. (CSTA, p. 13)

The new Dutch curriculum describes this implicitly as a *physical layer.*

"The candidate is able to explain the structure and functioning of digital artefacts through architectural elements, i.e., in terms of the physical, logical and application layer levels, and in terms of the components in these layers together with their interaction." (Netherlands 2016)

The high score on data by the old Dutch curriculum can be attributed to the learning objectives on *information systems, databases, relational schemas* and *query languages.*

"The candidate can name the elements of a relational schema and describe the significance of each element, and can convert information needs into a command formulated in a query language for a relational database. He can describe the features and aspects of database management systems, and name and use them for specific systems [...]" (Netherlands 2007, p. 3)

All these concepts are absent from the other four curricula. In the new Dutch curriculum, these concepts are treated in an elective theme on 'Databases'.

4.2   Algorithms

In this category, the documents differ in the amount of detail in which the learning objectives are described. We observed the CAS guidelines contains almost three times as many different concepts on algorithms as the French curriculum. The CAS guidelines, for example, explicitly states the notions of *sequence, selection* and *repetition.*

"- The idea of a program as a sequence of statements written in a programming language. - One or more mechanisms for selecting which statement sequence will be executed, based upon the value of some data item. - One or more mechanisms for repeating the execution of a sequence of statements, and using the value of some data item to control the number of times the sequence is repeated." (CAS, p. 14)

The CSTA guidelines go even further and, instead of *repetition* in general, explicitly specify *iteration* and *recursion.*

"Explain how sequence, selection, iteration, and recursion are building blocks of algorithms." (CSTA, p. 18)

Remarkably, the CSTA guidelines are the only curriculum specification in our sample that includes *recursion*. Likewise, CAS and the CSTA highlight the underlying notion of *instruction*, whereas France and the Netherlands do not.

> *"A computer program is a sequence of instructions written to perform a specified task with a computer."* (CAS, p. 14)

The new Dutch curriculum mentiones *instruction* only in the context of assembly languages.

> *"The candidate is able to write a simple program in a machine language, based on the description of an instruction set."* (Netherlands 2016)

We highlight some other concepts occurring in one single document.

Firstly, the explicit inclusion of *concurrency, parallelism* and *thread* in the CSTA guidelines is interesting. It is the only document to include these concepts.

> *"Describe the process of parallelization as it relates to problem solving."* (CSTA, p. 16)

> *"Demonstrate concurrency by separating processes into threads and dividing data into parallel streams."* (CSTA, p. 21)

Next, although *searching* and *sorting* appear in the French curriculum and the CAS and CSTA guidelines, the French curriculum is the only one of the three mentioning specific algorithms. It explicitly mentions *merge sort, breadth first search* and *depth first search*.

> *"Advanced algorithms: Merge sort; search for a path in a graph by a depth first search (DFS); finding a shortest path through a wide path (BFS). Skills: Understand and explain (orally or in writing) an algorithm. Questioning the effectiveness of an algorithm"* (France)

Finally, the French and new Dutch curriculum are the only ones including *state machines*.

> *"...describe a single event system with a finite state machine."* (France)

> *"The candidate is able to use finite automata for the characterization of certain algorithms."* (Netherlands 2016)

The old Dutch curriculum does not contain any concepts in the algorithm category. The new curriculum states objectives on the usage of *standard algorithms* and the *correctness* and *efficiency* of algorithms. It also provides an elective theme on 'Algorithms, Computability and Logic'.

4.3   Engineering

In the French curriculum, the category engineering has lower presence (4 %) compared to the other specifications. It does, however, contain pointers to *testing* and *verification*.

> *"Fixing a program: Test; instrumentation; error situations or bugs.*
> *Skills: Testing a developed program.* Optional: *using a development tool."*
> (France)

Testing and verification can be found in all other curricula, except for the old Dutch curriculum. The high score of the old Dutch curriculum in this category can be explained by the section on project management and related concepts like *specification, requirement, client* and *prototype*.

> *"The candidate can asses progress of a simple system development process, test prototypes, verify whether the final product meets the specifications of the client and evaluate whether the system meets the requirements."*
> (Netherlands 2007, p. 3)

Although the curriculum of the CSTA does not explicitly state concepts like *specification* and *requirement*, it does mention the *software development process* and *software life cycle* and the creation of problem statements in general.

> *"Describe a software development process used to solve software problems (e.g., design, coding, testing, verification)."* (CSTA, p. 18)

Furthermore, the CSTA standards have a strong focus on collaboration during software development. This is not surprising when we take the structure of the document into account. One of the five strands the document is built on is 'Collaboration' and a substantial part of the curriculum is dedicated to this strand. Concepts related to *teamwork* and *collaboration* are *peers, experts, pair programming, project teams, feedback, communication, feedback* and *socialization*. The CSTA document also mentions multiple productivity tools, development tools and collaboration tools explicitly.

> *"Use productivity technology tools (e.g., word processing, spreadsheet, presentation software) for individual and collaborative writing, communication, and publishing activities. Use collaborative tools to communicate with project team members (e.g., discussion threads, wikis, blogs, version control, etc.)."* (CSTA, p. 13)

The new Dutch curriculum contains a section dedicated to 'Informatics-specific skills', containing, amongst others, an objective on cooperation with(in) an inter- and intradisciplinary teams.

> *"The candidate is able to structurally cooperate in a team during the design and development of digital artefacts, and is able to cooperate with people from an application field."* (Netherlands 2016)

The inclusion of collaboration and tools make the CSTA guidelines and the new Dutch curriculum stand apart from the other specifications.

4.4   Society

A major difference between the CSTA guidelines and the new Dutch curriculum on one hand and the other specifications on the other, is the focus on computer science and society (both 11 %). One of the five strands of the CSTA guidelines is 'Society'. Therefore the subject covers a substantial part of the curriculum. The CAS curriculum contains a reference to *privacy*, whereas the French curriculum mentions *personal information* and *ownership*.

> *"Persistence of information: Data, including personal, may be stored for long periods without control by the persons concerned.*
> *Skills: Awareness of the persistence of information on digital networks. Understand the general principles to behave responsibly in relation to the rights of individuals in digital platforms."* (France)

The CSTA extends further with the inclusion of various other concepts ranging from *career perspectives*, via *software licenses* to *software piracy* and *legal behavior*.

> *"Exhibit legal and ethical behaviors when using information and technology and discuss the consequences of misuse."* (CSTA, p. 17)

The new Dutch curriculum contains objectives on 'Computer Science as a Perspective' and 'Ethical Conduct', both in computer science specific skills section. The curriculum includes a domain on 'Interaction', containing 'Social Aspects' and 'Privacy'. Furthermore, an elective theme on 'Social and Individual Influence of Informatics' contains concepts on social as well as legal influences of computer science on society.

> *"-The candidate is able to explain and predict the positive and negative effects of informatics and the networking society on the lives of individuals and on society. - The candidate is able to analyze legal aspects of the application of informatics in society. - The candidate is able to investigate the effects of technical, legal and social measures for privacy-related issues. - The candidate is able to reason about the influence of informatics on cultural expressions."* (Netherlands 2016)

4.5   Rest Category

The old Dutch curriculum contains a high number of concept occurrences in the rest category (8 %). This can be explained by the fact that the curriculum includes subjects on management and organization structures which are not mentioned in the ACM-IEEE body of knowledge. The old Dutch curriculum explicitly states students should know about project management and business structures, for example in the following learning objective taken from the domain 'Basic Concepts and Techniques'.

"The candidate knows the overall organizational structure of companies. He knows the characteristics of a project and can indicate why, during major changes in a information system of a company, one often chooses to use a project." (Netherlands 2007, p. 2)

No other document, including the revised Dutch curriculum, contains these kind of 'contextual' objectives.

5 Conclusion and Discussion

Our analysis suggests that there are similarities between recent curriculum specifications with respect to a number of knowledge categories such as algorithms and data.

We also found differences with respect to emphasis. The French curriculum specification appears to have a stronger emphasis on data. In the new Dutch curriculum and the CSTA standards, concepts around engineering receive substantial attention. The new Dutch curriculum and the CSTA guidelines appear to stress societal aspects more than the other documents in our sample. Software engineering and social and ethical topics cover together a quarter of the concepts occurrences in the CSTA standards. We found the most specific descriptions with respect to algorithms in the CAS guidelines and in the French curriculum. The CSTA standards are more outspoken on collaboration in the context of engineering.

Compared to the old curriculum, the new Dutch core curriculum appears to have smaller emphasis on hardware (architecture, networking) in favor of software and engineering content.

Our method appeared very useful to compare curriculum documents that differ considerably with respect to description style and level of detail. Moreover, the combination of a quantitative and qualitative investigation turned out to be valuable.

Our findings are consistent with those found for K–9 guidelines (Barendsen et al. 2015), but some are more outspoken, for example with respect to algorithms.

The categorization in our study has more categories than the ACM classification used by Hubwieser et al. (2011) and appears to give a more balanced distribution of conceptual content.

This paper contains a selection of observations; the full analysis was instrumental during the construction of the new Dutch curriculum. Our small-scale study spanned a limited number of documents. We plan to apply our method to a larger collection of curriculum documents and guidelines.

The stepwise procedure in this small-scale pilot study made it possible to code an entire document in a reasonable amount of time. The authors quickly reached consensus concerning coding differences. Throughout the process, the intercoder agreement appeared high. We plan to analyze reliability of the method in a more quantitative way.

It will be interesting to investigate whether conceptual differences between curricula can be related to other characteristics, such as the global curriculum intention in the sense of the 'goals' described by Biesta (2015).

References

Académie des Sciences: L'enseignement de l'informatique en France: Il est urgent de ne plus attendre. http://www.academie-sciences.fr/pdf/rapport/rads_0513.pdf. Accessed Apr 2016

ACM/IEEE-CS Joint Task Force on Computing Curricula: Computer science curricula 2013 (Technical report). ACM Press and IEEE Computer Society Press, December 2013

Barendsen, E., Fisser, P., Krüger, J., Tolboom, J.: Herziening van het Nederlandse informaticacurriculum havo-vwo. Paper presented at ORD 2014, Groningen (2014)

Barendsen, E., Mannila, L., Demo, B., Grgurina, N., Izu, C., Mirolo, C., Stupurienė, G.: Concepts in K–9 computer science education. In: Proceedings of the 2015 ITiCSE on Working Group Reports, pp. 85–116. ACM (2015)

Barendsen, E., Tolboom, J.: Advies examenprogramma informatica vwo-havo: inhouden invoering. SLO, Enschede (2016)

Biesta, G.J.J.: Good Education in an Age of Measurement: Ethics, Politics, Democracy. Routledge, Abingdon (2015)

British Department for Education: Computing programmes of study: key stages 1 and 2. National curriculum in England (2013). http://www.computingatschool.org.uk

Cohen, L., Manion, L., Morrison, K.: Research Methods in Education. Routledge, London, New York (2013)

Computing at School Working Group: Computer science: a curriculum for schools (2012). http://www.computingatschool.org.uk/data/uploads/ComputingCurric.pdf. Accessed Sep 2013

CSTA: K-12 computer science standards. ACM (2011). http://csta.acm.org/Curriculum/sub/K12Standards.html

Dagiene, V., Jevsikova, T., Schulte, C., Sentance, S., Thota, N., et al.: A comparison of current trends within computer science teaching in school in Germany and the UK. In: Informatics in Schools: Proceedings of the 6th International Conference ISSEP 2013—selected papers, pp. 63–75 (2013)

Furber, S.: Shut down or restart? The way forward for computing in UK schools. The Royal Society, London (2012)

Gander, W., Petit, A., Berry, G., Demo, B., Vahrenhold, J., McGettrick, A., Meyer, B.: Informatics education: Europe cannot afford to miss the boat. (Report of the Joint Informatics Europe & ACM Europe Working Group on Informatics Education) (2013). http://www.informatics-europe.org/images/documents/informatics-education-europe-report.pdf. Accessed Aug 2013

Grgurina, N., Tolboom, J.: The first decade of informatics in Dutch high schools. Inf. Educ. **7**(1), 55–74 (2008)

Hubwieser, P.: The Darmstadt model: a first step towards a research framework for computer science education in schools. In: International Conference on Informatics in Schools: Situation, Evolution, and Perspectives, pp. 1–14 (2013)

Hubwieser, P., Armoni, M., Brinda, T., Dagienė, V., Diethelm, I., Giannakos, M.N., Schubert, S.: Computer science/informatics in secondary education. In: Proceedings of the 16th Annual Conference Reports on Innovation and Technology in Computer Science Education-Working Group Reports, pp. 19–38 (2011)

Kaczmarczyk, L., Dopplick, R.: Preparing students for computing workforce needs in the US. ACM SIGCSE Bull. **46**(2), 8 (2014)

KNAW: Digitale geletterdheid in het voortgezet onderwijs: vaardigheden en attitudes voor de 21ste eeuw. Koninklijke Nederlandse Akademie van Wetenschappen, Amsterdam (2012)

Ministère de l'Éducation Nationale: Enseignement de spécialité d'informatique et sciences du numérique de la série scientifique – classe terminale (2012). http://www.education.gouv.fr/pid25535/bulletin_officiel.html?cid_bo=57572

Samaey, G., Van Remortel, J., Bersini, H., Bruynseraede, Y., Dekelver, J., Laender, F.D., Wyffels, F.: Informaticawetenschappen in het leerplichtonderwijs. Koninklijke Vlaamse Academie van België voor Wetenschappen en Kunsten, Brussel (2014)

Schmidt, V.: Handreiking schoolexamen informatica havo/vwo. SLO, Enschede (2007)

Steenvoorden, T.: Characterizing fundamental ideas in international computer science curricula (unpublished Master's thesis). Radboud University, The Netherlands (2015)

It's Computational Thinking! Bebras Tasks in the Curriculum

Valentina Dagienė[1] and Sue Sentance[2(✉)]

[1] Vilnius University Institute of Mathematics and Informatics,
Akademijos Street 4, 08663 Vilnius, Lithuania
valentina.dagiene@mii.vu.lt
[2] Department of Education and Professional Studies, King's College London,
150 Stamford Street, London SE1 9NH, UK
sue.sentance@kcl.ac.uk

Abstract. Bebras is an award-winning, international contest and challenge in informatics that has been running for 12 years in primary and secondary schools, with 50 countries now participating. From a single contest-focused annual event the Bebras developed to a multifunctional challenge; an activities-based educational community-building network has grown up where the development of Bebras tasks has taken a very significant role. Bebras tasks present a motivating way to introduce computer science concepts to students as well as developing computational thinking skills. Tasks are categorized in terms of the concepts being covered, and each task includes an explanation of how the task relates to informatics. In this paper we propose that Bebras tasks can be used within the school curriculum (whether it is called informatics, computer science, computing or information technology) to promote computational thinking and provide teaching materials. We give examples of Bebras tasks that could be incorporated into the curriculum, and make recommendations for schools wishing to develop children's computational thinking skills.

Keywords: Bebras contest · Computational thinking · Computer science education · Informatics curriculum · Informatics education · Task solving

1 Introduction

There is an increasing focus on computational thinking within the teaching of computer science, computing or informatics (from here on referred to as informatics) in school. Computational thinking was only recently popularised as a concept in 2006 by Wing (2006), although the original definition stems from Papert (1996). Wing claims that computational thinking is for everyone and involves "solving problems, designing systems and understanding human behaviour, by drawing on the concepts fundamental to computer science" (Wing 2006, p. 34). Some new informatics curricula have a significant focus on computational thinking skills being developed, for example in England (Brown et al. 2014) and Poland (Syslo and Kwiatkowska 2015). In the longstanding Bebras contest (Bebras 2016), tasks are designed which demonstrate computer science principles whilst engaging students in problem-solving in a motivating way.

© Springer International Publishing AG 2016
A. Brodnik and F. Tort (Eds.): ISSEP 2016, LNCS 9973, pp. 28–39, 2016.
DOI: 10.1007/978-3-319-46747-4_3

Bebras is an informatics education community-building model and is designed to promote informatics learning in school by solving short concept-based tasks (Dagiene and Stupuriene 2016). Tasks are the most important component of the Bebras model. Each Bebras task should include at least one informatics concept, attract children's attention by a story, picture or interactivity, be short (fits in a computer screen), and not require specific technical knowledge. Some countries use the Bebras to strengthen collaborative learning; for example, in Germany pupils solve Bebras tasks in pairs during a contest and discussions are allowed between the pairs.

Alongside the initial goal of the Bebras project to motivate pupils to be more interested in informatics topics there is a strong intention to deepen algorithmic, logical and operational thinking and, more recently, computational thinking as well. The Bebras challenge intends to promote students' interest in informatics (also in a better understanding of the usage of technology) from the very beginning at school and to motivate students to learn and master technology (Dagiene and Futschek 2008). In the past few years, the number of Bebras challenge participants has been notably growing and exceeded 1.3 million during the Bebras week in November 2015.

In this paper we argue that Bebras is thus a non-formal activity and a possible way in which to incorporate computational thinking into the primary and secondary school curricula, and suggest some exemplar activities to incorporate this.

2 Computational Thinking

The term 'computational thinking' is primarily accredited to Jeanette Wing (Wing 2006), but actually originated with Seymour Papert (Papert 1996). There are differences between these two definitions in that Wing's definition is more focused on problem solving and Papert's definition is more focussed on ideas and analysis (Mannila et al. 2014). Subsequent research has expanded and interpreted the term further (Lu and Fletcher 2009; Grover and Pea 2013; Selby and Woollard 2013).

Computational thinking is not entirely embraced by all; critics suggest that the term is narrowing (Denning 2009) or that computational thinking processes are widespread in other sciences (Hemmendinger 2010). Among other contributions coming from educators, Lee et al. (2011) suggest that we should start from practical examples of what we mean by computational thinking, and identify the terms "abstraction", "automation", and "analysis" as being particularly useful to understand how young pupils can deal with novel problems. Indeed, there is a huge interest in computational thinking as a means of explaining the thinking processes in informatics in school education (K-12); in USA computational thinking underlies the new curricular developments of the Computer Science Teacher Association in USA (CSTA) and Code.org; in England, computational thinking is at the core of a mandatory new Computing curriculum from age 5 until 16 (Department for Education 2013); and Google have launched a teacher development MOOC purely around computational thinking (Google 2016). Attention has turned to the identification of a set of skills that can be seen to comprise a broad definition of computational thinking, and that encompass the logical and problem-solving skills and thought processes that are applied by computer scientists in their work.

The work by Computing At School in the UK defines the five key computational thinking skills used in K-12 as abstraction, decomposition, algorithmic thinking, evaluation and generalisation (Csizmadia et al. 2015). There is also the question of how much computational thinking development is around computer programming and related topics, for example, physical computing (Przybylla and Romeike 2014). Lu and Fletcher 2009 take the view that computational thinking can be separated from programming, and should be taught before programming teaching starts. In addition, Wing's definition of computational thinking includes understanding the consequences of scale, not only for reasons of efficiency but also for economic and social reasons. CSTA in USA adds broader attitudes like the ability to deal with complexity and open-ended problems, tolerance for ambiguity, and ability to work with others to achieve a common goal (ISTE&CSTA 2011).

Computational thinking is explicitly mentioned in some curricular, for example, here in the curriculum in England, referring to pupils aged 7–11: "*Pupils should be taught to: ... Solve problems by decomposing them into smaller parts*" (Department for Education 2013).

3   Computational Thinking and Bebras

One of the drivers of the Bebras community is a shared understanding that learning concepts at an early age is important for a deeper understanding of various informatics topics. The Bebras learning model focuses on informatics concepts by supporting an understanding of computer science phenomena and the development of computational thinking. For the purposes of Bebras we adopt the broad view that computational thinking is a problem-solving process that includes (but is not limited to) the following characteristics (ISTE&CSTA 2011):

- Formulating problems in a way that enables us to use a computer and other tools to help solve them.
- Logic and predicting analytics.
- Data organizing and analysing.
- Representing data through abstractions such as models and simulations.
- Automating solutions through algorithmic thinking (a series of ordered steps).
- Identifying, analysing, and implementing possible solutions with the goal of achieving the most efficient and effective combination of steps and resources.
- Generalizing and transferring this problem solving process to a wide variety of problems.

One suggested classification of computational thinking skills follows the work of Selby and Woollard (2013) and has been adopted by Computing At School in the UK in developing guidance on computational thinking for teachers (Csizmadia et al. 2015). This describes aspects of computational thinking skills exhibited by students as falling into the five categories below:

1. Abstraction
2. Algorithmic thinking

3. Decomposition
4. Evaluation
5. Generalisation

Based on a previous Bebras categorisation system (Dagiene and Futschek 2008) and further developments with relation to Bebras tasks' content, we can identify the main informatics concept introduced in the task and very broadly divide the content of the task into one of these five areas (categories):

1. Algorithms and programming
2. Data, data structures and representations (includes graphs, data mining)
3. Computer architecture and processes (includes anything to do with how the computer works - scheduling, parallel processing)
4. Communications and networking (includes cryptography, cloud computing)
5. Interaction (Human-Computer Interaction, HCI), systems and society

Analyses of the Bebras tasks used in the 2014 contest were conducted according to the cognitive skills' domains (Bloom taxonomy): this showed that the most tasks demonstrated higher-order thinking skills in the Bloom's taxonomy: Understanding, Applying, Analysing and Evaluating (Dagiene and Stupuriene 2014). In another analysis examining the topics of all Bebras tasks used between 2010 and 2014, the most commonly occurring computational thinking topics were algorithms (66 %) and data representation (38 %), followed by abstraction (16 %) (Barendsen et al. 2015).

In this paper we analyse Bebras tasks that were chosen by Lithuania and UK for all age groups in 2015: in total these amount to 52 tasks, of which the two countries have 35 in common (presented in italics). For each task we allocated the primary and most important computational thinking skill being developed in that task (Table 1), even though we acknowledge that a given task may in some cases develop more than one computational thinking skill.

In Table 1 we can see that of the 52 tasks chosen between the two countries, 22 of them involved some degree of algorithmic thinking in finding a solution. 11 tasks involve the skill of evaluation, 8 demonstrate abstraction, 6 decomposition, and 5 generalisation. Tasks can demonstrate more than one computational thinking skill but in this instance we have highlighted the most dominant one. The emphasis on algorithmic thinking (42 % of tasks) is interesting and supports the observations by Barendsen et al. (2015) about previous tasks. Is it the case that computer scientists use algorithmic thinking more than other computational thinking skills? Or do Bebras task authors find it easier to write tasks that involve either executing, debugging or creating an algorithm? We surmise that it may be a combination of these factors: Bebras tasks are short and designed to be solved within 3 min. It may be difficult to generate tasks that demonstrate a lot of decomposition or evaluation in a short task. However, a key aspect of computer science at school level is the design and execution of algorithms, which supports the development of programming skills, so it may not be surprising that so many algorithmic thinking tasks make their way into the Bebras contest.

Table 1. Bebras 2015 task analysis according to computational thinking (CT) skills

CT Skill	Tasks	Example
Abstraction	*Beaver the Alchemist* Busy Beaver *Drawing Stars* Fried egg *Geocaching* *Popularity* Trains *Walnut Animals*	**Walnut animals:** With walnut animals, we abstract from features like fur and size. We represent the animal only by the structure of its body; the rest is unimportant. This structure is preserved even when the animals are transformed. A computer scientist must recognise what is important, what can be left out, and how structures are similar
Algorithmic thinking	*Beaver Logs* *Biber Hotel* *Bowl Factory* *Building a Chip* Button Game Car Transportation *Chakhokhbili* *Crane operating* *Cross Country* Decorating Chocolate *Drawing Patterns* *Dream Dress* *Fair Share* *Irrigation system* Left Turn! *Mushrooms* Pencils Alignment Reaching the Target *Supper Power Family* *Theatre* *Throw the Dice* *You Won't Find It*	**Biber hotel:** The structure of the beaver hotel is a so-called "binary tree", meaning that from every there are two branches leaving to further rooms. The room number facilitates further navigation. Data on a computer can also be organised in such a way. Despite having several millions of entries, an entry (or its absence) can be found in less than 25 comparisons. In fact, with at most n comparisons it is possible to distinguish between 2^n-1 entries **Crane operating:** In this task a sequence of instructions is searched for. Two objects can only be changed if one of the objects is placed at an empty place. Most computers still work with sequentially-run programs, so each exchange operation in the memory of the computer also needs an extra space

(Continued)

Table 1. (*Continued*)

CT Skill	Tasks	Example

Decomposition	*Animation*
	Fireworks
	Pirate Hunters
	Stack
	  Computer
	Quick Beaver
	  Code
	Word Chains
		Stack computer: The usual notation for arithmetic expressions is not the easiest to understand for a computer, or rather, it takes a more complicated program to process such expressions. However writing a program to analyse expressions in postfix notation (or stack computer) is much easier. To solve this task the expression must be broken down (decomposed) into its individual parts

Evaluation	*Animal*
	  Competition
	Beaver Gates
	Beaver
	Tutorials
	Birds
	Bracelet
	Birthday
	Balloons
	Data
	  Protection
	Email Scam
	Robot the
	  Stairs
	Setting the
	  Table
	Turn the Cards
		Bracelet: It is important to be able to recognise patterns which may be useful to us. Recognising patterns helps us to find similarities in things that may look different at first, but have something in common. This task also deals with verifying a proposed solution: the possible answers need to be checked against the original bracelet to see if they meet the required order of the shapes

Generalisation	*Beaver Lunch* Kangaroo *Mobiles* RAID Array Spies	**Mobiles:** If you detach a stick (except the uppermost one) from a mobile, you have a mobile again, with the detached stick being the uppermost stick now. That is, the parts of a mobile are constructed in the same way as the full mobile is constructed. If a single figure is considered as a mobile, mobiles may be defined as follows: a mobile is either (a) a single figure, or (b) a stick with one or more mobiles attached to it. In order to define a "mobile", we use the term "mobile" itself. That is a recursive definition, an important concept in computational thinking

4 Bringing Bebras into the Curriculum

As seen above, there is a clear link between Bebras tasks and the development of computational thinking skills, thus demonstrating their potential to be used in the curriculum to develop these skills. In addition, Bebras tasks can be used to demonstrate

specific informatics topics and concepts. In this section, we will illustrate this with some examples of previous Bebras tasks that could be incorporated into an Informatics curriculum in any country. Three curriculum areas have been selected that are currently taught in schools in England and Lithuania, together with some Bebras example tasks are that can be used in school; these areas are: data structures, logical operators and networks.

4.1   Learning About Data Structures

There have been many Bebras tasks in previous years that could be introduced to students which might support an understanding of data structures such as trees, graphs, stacks queues etc. Two examples are discussed below (Figs. 1 and 2).

The structure of the beaver den is a so-called "binary tree", meaning that from every room (a node) there are (possibly) two branches leaving to further rooms. The room-number (or any other ordered data) serves to navigate and find a room again. Data on a computer can also be organised in such way (like for instance names and phone numbers). In fact, with at most n comparisons (depth of the tree) it is possible to distinguish between 2^n-1 entries. For n = 10 we have 1023 possible entries, for n = 20 we have a little over 1 million entries and for n = 30 over one billion.

Over the years, the beavers constructed a huge beaver den with many, many rooms. The rooms are arranged in a particular tunnel structure and numbered.

Click on the picture to move through the den. Find the room with **number 1337**.

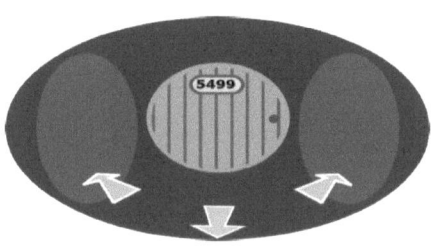

Fig. 1. Biber hotel: a task on a binary tree concept (Ivo Blöchliger, Switzerland)

The Animation task shown in Fig. 2 deals with a data structure concept, in particular that of class, which is very important concept in object oriented programming.

B-taro is planning an animation, which shows a sequence of pictures of a face. The animation should run smoothly. The order of the pictures will be correct if only one attribute of the face changes from one picture to the next. Unfortunately, the pictures got mixed up. Now B-taro must find the correct order again. Luckily, he knows which picture is last. He labels the five other pictures with letters A to E.

In order to find the differences between the pictures, pupils have to find out about the essential attributes of the depicted faces first. The list of attributes and their possible values is: ears: small, large; mouth: plain, smile; nose: small, large; number of teeth: 2, 3; whiskers: curly, straight. For instance, pupils can describe the first face as a list of attribute-value pairs: (ears: small; mouth: plain; nose: large; number of teeth: 3; whiskers: straight).

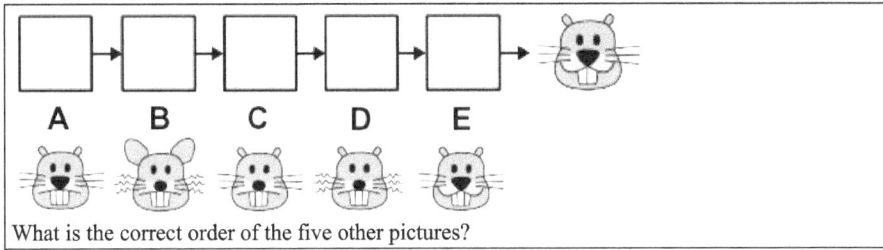

What is the correct order of the five other pictures?

Fig. 2. Animation – a task on a class concept of object-oriented programming (Tomohiro Nishida, Japan))

4.2   Learning About Logical Operations

In many countries, understanding logical operations is a key part of the informatics curriculum. In the national curriculum in England, pupils have to "understand simple Boolean logic [for example, AND, OR and NOT] and some of its uses in circuits and programming" at ages 11–14 (Department for Education 2013). Bebras tasks can be focused around different aspects of this topic, particularly tasks where students have to demonstrate an understanding of AND, OR and NOT, or combinations of these operations, in order to solve a task (Fig. 3). The use of such tasks can have a direct applicability to the curriculum.

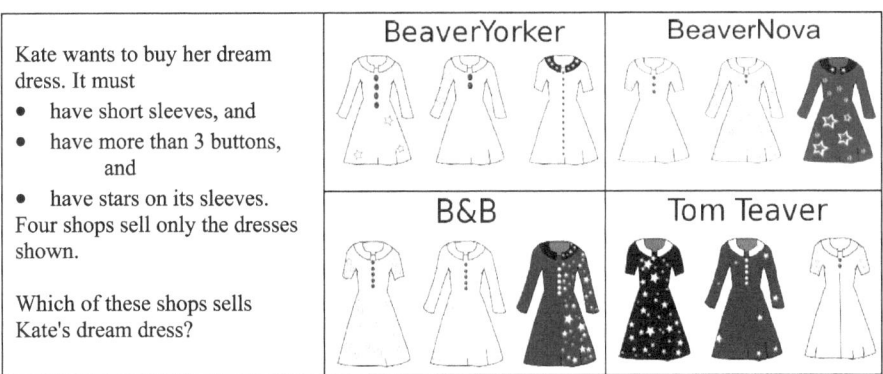

Fig. 3. Dream dress (Karolína Mayerová, Slovakia)

The Dream Dress task involves statements (conditions) that must be evaluated (determined to be true or false) for a set of objects (coats). Conditions and their evaluation is an important part of programming and algorithmic thinking. Conditions can be simple statements. However, more complex statements can be formed using logical operators such as AND, OR, NOT, etc. This task uses the AND operator.

4.3 Learning About Networks

The topic of networks is very broad; it can be found in various forms in many countries' informatics curricular (Barendsen et al. 2015). At school level, this topic could cover topologies, communication, networking protocols, security and the way that the internet is structured. The communication offered by networking can also be seen in examples of social networks, as in the following task (Fig. 4).

A social network is a network used for communication and will be familiar to many students engaging in the Bebras contest. Social networks present us with examples of large and complex networks. It is not always obvious that by posting something on a friend's page, it might be available to people other than the close friend.

Social networks themselves are incredibly powerful tools in today's world. Computing statistics on their users and their pages is useful to marketing departments and anyone else trying to understand a person or group of people. *Instadam* could also be interpreted as a model of a miniature internet, with the beavers being websites and friends as pages "linked to". Search engines typically rank these websites by some measure of popularity or importance, at least by the number of links to and from the website. A widely used way to find the result by using a computer is to use the flood fill algorithm which can cope with systems with more than the two iterations in this example.

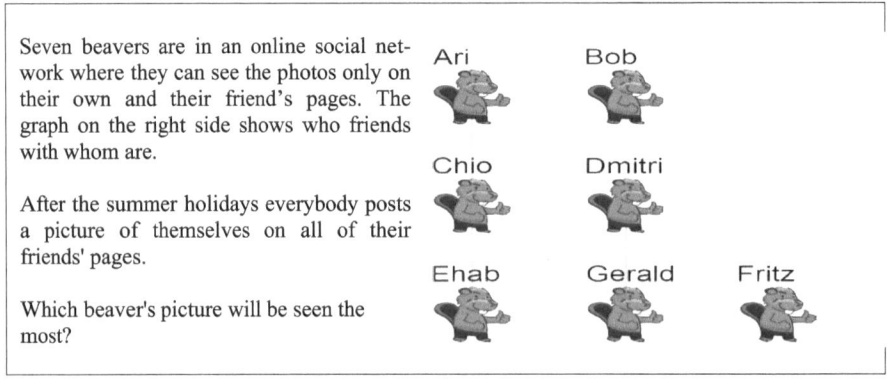

Seven beavers are in an online social network where they can see the photos only on their own and their friend's pages. The graph on the right side shows who friends with whom are.

After the summer holidays everybody posts a picture of themselves on all of their friends' pages.

Which beaver's picture will be seen the most?

Ari   Bob   Chio   Dmitri   Ehab   Gerald   Fritz

Fig. 4. Popularity (J.P. Pretti, Canada, Cristian Datzko, Switzerland, Sarah Hobson, Australia)

Another key aspect of networks which will be covered in the school curriculum is security. The example Spies (Fig. 5), focusing on spies exchanging information, illustrates a Bebras way of introducing this in school.

These examples illustrate the direct connection from topic to task which can be exploited in the classroom. All examples given here are from the 2015 contest, but as the competition has run since 2004, there are many more examples of tasks that demonstrate computer architecture, principles of operating systems, cryptography and other concepts relevant to the curriculum.

Every Friday, six spies exchange all the information they've gathered in the week. A spy can never be seen with more than one other spy at the same time. So they have to conduct several rounds of meetings where they meet up in pairs and share all information they have at that point. The group of 6 spies needs only three rounds to distribute all secrets: Before the meetings each spy holds a single piece of information. (spy 1 knows 'a', spy 2 knows 'b', etc.). In the first round spies 1 and 2 meet and exchange information so now both know 'ab'. The diagram shows which spies meet in each round with a line. It also shows which pieces of information they all have. After three rounds all information has been distributed.

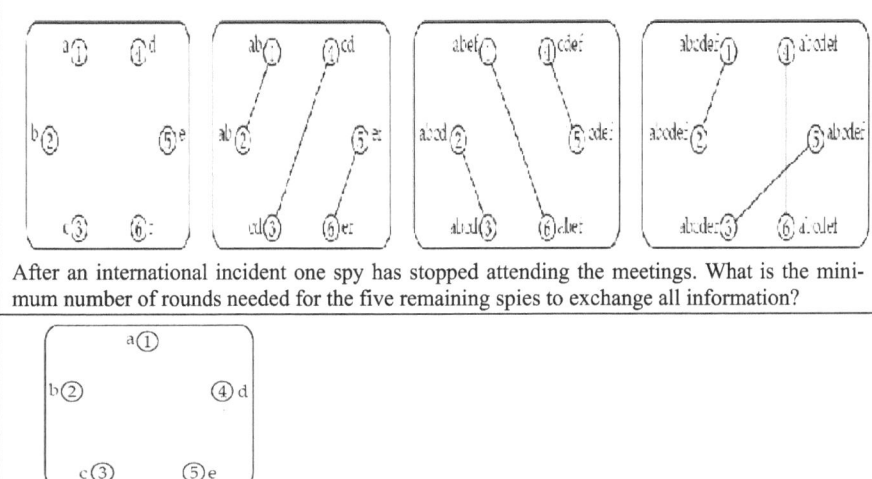

After an international incident one spy has stopped attending the meetings. What is the minimum number of rounds needed for the five remaining spies to exchange all information?

Fig. 5. Spies (Janez Demsar, Slovenia)

5 Pedagogical Issues

The question remains as to the identification of teaching approaches that can draw on Bebras tasks as a resource. To a certain extent the country's curriculum will dictate which tasks are appropriate to be incorporated into a scheme of work. However the tasks lend themselves to being interesting starter tasks for the beginning of a lesson or plenary tasks, for the formative assessment part of a lesson. Currently many teachers use previous tasks as preparation for their students prior to the contest each November; with the growing number of available tasks Bebras tasks could be used in teaching all year round.

Planning lessons around relevant Bebras tasks can only be achieved if Bebras tasks are available and the content is clearly signposted. A new two-dimension categorisation system being proposed for Bebras tasks (Dagiene and Sentance in review) will assist with this. Within this categorisation, each task is classified in terms of its computational thinking skills and informatics concepts. Teachers will be able to use this categorisation to select material for teaching. One situation that can be envisaged is that each country (or countries sharing a common language) has a database of previous tasks that could be searched via concept or computational thinking skill.

Another key area for consideration is assessment. In the Bebras contest, tasks are marked automatically and teachers have access to the final results of their students. By using the tasks for formative assessment in lessons, teachers can track their students' progress in developing computational thinking skills.

6 Conclusion

Bebras tasks present a motivating way to introduce informatics concepts to students as well as developing computational thinking skills. Bebras task developers seek to choose interesting tasks (problems) for enabling students to understand informatics and to think deeper about technology. Moving forward these tasks should cover a range of as many different informatics topics as possible. In addition tasks can be designed which aid the development of core computational thinking skills such as abstraction, algorithmic thinking, decomposition, evaluation and generalisation.

In this paper, the use of Bebras tasks in teaching to promote computational thinking and the introduction of concepts has been suggested through possible examples. Bebras tasks are categorized in terms of the concepts being covered, and can also include a categorisation by computational thinking skill. To support teachers developing lessons, each task includes an explanation of how the task relates to informatics. This can also support teachers who are not fully confident in the subject matter around the tasks, and add to their own professional development. Further work is needed to evaluate the extent to which the use of these tasks in the classroom can support the learning and assessment of computational thinking.

Acknowledgements. The authors thank all members of the international Bebras community who took part in task development and in this way influenced the outcomes of this paper. In addition, we are grateful to Chris Roffey for the development of the UK Bebras Answer Booklet 2015 from which we have taken some ideas for explanation of the example tasks in this paper.

References

Aarts, R.M.: Gossiping. From *MathWorld*–A Wolfram Web Resource. Created by Weisstein, E.W. (2016). http://mathworld.wolfram.com/Gossiping.html. Accessed 30 Apr 2016

Barendsen, E., Manilla, L., Demo, B., Izu, C., Grugina, N., Mirono, C., Sentance, S., Settle, A., Stupuriene, G.: K-9 concepts in computer science education. ITICSE Working Group report (2015)

Bebras International Challenge on Informatics and Computational Thinking. http://www.bebras. org/en/facts. Accessed 30 Apr 2016

Brown, N., Sentance, S., Crick, T., Humphreys, S.: Restart: the resurgence of computer science in UK schools. ACM Trans. Comput. Educ. **14**(2), 9 (2014)

Csizmadia, A., Curzon, P., Dorling, M., Humphreys, S., Ng, T., Selby, C., Woollard, J.: Computational Thinking: A Guide for Teachers (2015). http://computingatschool.org.uk/computationalthinking. Accessed 10 Apr 2016

Dagienė, V., Futschek, G.: Bebras international contest on informatics and computer literacy: criteria for good tasks. In: Mittermeir, R.T., Sysło, M.M. (eds.) ISSEP 2008. LNCS, vol. 5090, pp. 19–30. Springer, Heidelberg (2008)

Dagiene, V., Sentance, S.: Computational thinking and the Bebras challenge: developing a new task categorization system (in review)

Dagiene, V., Stupuriene, G.: Informatics education based on solving attractive tasks through a contest. Commentarii informaticae didacticae 7, 97–115 (2014)

Dagiene, V., Stupuriene, G.: Bebras - a sustainable community building model for the concept based learning of informatics and computational thinking. Inform. Educ. 15(1), 25–44 (2016)

Denning, P.J.: Beyond computational thinking. Commun. ACM 52(6), 28–30 (2009)

Department for Education: The National Curriculum in England: Computing Programmes of Study (2013). https://www.gov.uk/government/publications/national-curriculum-in-england-computing-programmes-of-study. Accessed 30 Apr 2016

Google for Educators: Exploring Computational Thinking (2016). https://www.google.com/edu/resources/programs/exploring-computational-thinking. Accessed 30 Apr 2016

Grover, S., Pea, R.: Using a discourse-intensive pedagogy and Android's App Inventor for introducing computational concepts to middle school students. In: Proceedings of 44th SIGCSE Technical Symposium on Computer Science Education, pp. 723–228. ACM (2013)

Hemmendinger, D.: A plea for modesty. ACM Inroads 1(2), 4–7 (2010)

ISTE&CSTA (International Society for Technology in Education & the Computer Science Teachers Association): Operational definition of computational thinking for K-12 education (2011). https://csta.acm.org/Curriculum/sub/CurrFiles/CompThinkingFlyer.pdf

Lee, I., Martin, F., Denner, J., Coulter, B., Allan, W., Erickson, J., Malyn-Smith, J., Werner, L.: Computational thinking for youth in practice. ACM Inroads 2(1), 32–37 (2011)

Lu, J.J., Fletcher, G.H.: Thinking about computational thinking. ACM SIGCSE Bull. 41(1), 260–264 (2009)

Mannila, L., Dagiene, V., Demo, B., Grgurina, N., Mirolo, C., Rolandsson, L., Settle, A.: Computational thinking in K-9 education. In: Proceedings of Working Group Reports of the 2014 on Innovation & Technology in Computer Science Education Conference, ITiCSE-WGR, pp. 1–29. ACM, New York (2014)

Papert, S.: An exploration in the space of mathematics educations. Int. J. Comput. Math. Learn. 1, 95–123 (1996)

Przybylla, M., Romeike, R.: Physical computing and its scope - towards a constructionist computer science curriculum with physical computing. Inform. Educ. 13(2), 225–240 (2014)

Selby, C., Woollard, J.: Computational thinking: the developing definition (2013). http://eprints.soton.ac.uk/356481. Accessed 30 Apr 2016

Syslo, M.M., Kwiatkowska, A.B.: Introducing a new computer science curriculum for all school levels in Poland. In: Brodnik, A., Vahrenhold, J. (eds.) ISSEP 2015. LNCS, vol. 9378, pp. 141–154. Springer, Heidelberg (2015). doi:10.1007/978-3-319-25396-1_13

Wing, J.M.: Computational thinking. Commun. ACM 49(3), 33–35 (2006)

How to Attract the Girls: Gender-Specific Performance and Motivation in the Bebras Challenge

Peter Hubwieser[✉], Elena Hubwieser, and Dorothee Graswald

Technical University of Munich, TUM School of Education,
Arcisstrasse 21, 80333 Munich, Germany
{peter.hubwieser,elena.hubwieser,dorothee.graswald}@tum.de
http://www.ddi.tum.de

Abstract. Potentially, the international Bebras Challenge could provide a facility to arouse the enthusiasm of girls for Computer Science. Yet, its effect will depend on the personal success in the challenge, which is likely to correlate with personal motivation. Thus, we have compared the performance of the girls and the boys in the German Bebras challenge of 2014. Overall, the boys were more successful. The differences increase dramatically with the age of the participants. Additionally, we have compared the average performance of boys and girls in every task. It turned out that girls performed better in a certain task, if three conditions of Kellers ARCS model of motivation are met: the tasks have to look nice to attract the attention of the girls, they have to represent a situation relevant for real life, and they have to be comparably easy to solve.

Keywords: Bebras challenge · Gender differences · Informatics · Computer science · Motivation · Self-efficacy

1 Introduction

The attempt to engage more women in Computer Science (CS) has turned out to be a substantial challenge over many years in many countries, for example in Germany [24]. As one of the reasons was found that in our society woman regard themselves to be weaker performing in Mathematics and Computer Science compared to men [20]. Due to the obvious urgency of this problem, over the last decades many projects have been launched to motivate women to engage in Computer Science. Yet, as already very young girls seem to have different attitudes toward CS compared to even-aged boys [8], all attempts to influence adult women might come too late. In this regard, the international Bebras challenge provides a dawn of hope, as it aims to raise the interest and motivation in CS, addressing children at the age of 10 already. Yet, as it has turned out that intrinsic motivation requires the experience of competence [12], the challenge will have a positive impact on girls only if they are able to solve a satisfying number of tasks. To find out if this is the case and to detect differences between

© Springer International Publishing AG 2016
A. Brodnik and F. Tort (Eds.): ISSEP 2016, LNCS 9973, pp. 40–52, 2016.
DOI: 10.1007/978-3-319-46747-4_4

boys and girls, the outcomes of the challenge have to be analyzed regarding the gender of the participants. In certain tasks of the German Bebras challenge of 2009, we have already found significant differences between boys and girls [14]. In this paper, we compare the performance of boys and girls in the German Bebras challenge of 2014. In average, the girls had a significantly lower performance compared to the boys. Yet, in the two younger age groups, the girls outperformed the boys in certain tasks. The analysis of these tasks demonstrates that girls can be motivated by the first three factors of Kellers ARCS Model [17]: *Attention, Relevance and Confidence.*

2   Theoretical Background

The concept of self-efficacy describes the self-assessment of a person regarding his/her ability to master certain tasks or to reach certain goals [2]. A persons' self-efficacy also seems to represent a reasonable predictor for his/her efforts to master these tasks, because people tend to estimate their future performance according to past successes or failures [3]. It has been found that women assess their self-efficacy in general lower than men on the same performance level [19]. The different working practices of young people were investigated for example by Lamoureux, Beheshti, Abuhimed and AlGhamdi [19]. They observed the behavior of their probands during the work on a project and found that girls were less confident. On the other hand, in case of problems, young men were found to give up earlier. In total, the girls worked more carefully and purposefully and also showed more satisfaction at the end of the project.

A study of Cen, Ruta, Powell and Ng [7] investigated the collaboration in single-sex groups. The female groups showed clearly more collaboration and achieved better results than the male groups. Hubwieser and Mühling have confirmed this effect in the context of the German Bebras Challenge of 2009, restricting their analysis to a set of tasks that seem to require spatial intelligence only [14].

To explain the constantly low portion of young women among computer science students, numerous studies on gender differences have been conducted, see [23,24]. It turned out that certain stereotypes regarding computer science have established [15]. For example, Graham and Latulipe reported that girls regard programmers as "geeky" [13]. On the other hand, many women lack confidence to master a computer science course due to their low perception of self-efficacy. Additionally, they regard themselves as gifted only in a few partial fields of computer science. In consequence, they do not enroll in such courses or drop out quite early [5,20]. Even after receiving a computer science degree, stereotypes and prejudices form barriers for women. Often they are simply not regarded as evenly competent as their male colleagues [22]. In addition, it has been reported that they tend to underestimate their skills, even if they have the same or better grades compared to their male fellows [5,21]. Boys seem to be more interested in Computer Science because they like problem solving [23]. They work on problems that can be solved on a computer more often and more intensively compared to women. By this way, they seem to learn certain technical skills better, which provides additional advantages compared to girls, who generally seem to start their studies in Computer Science with less prior experience [21]. Furthermore, men often do programming just for itself and therefore work more frequently in a trial and error method [16]. Out of these experiences, they tend to have a more relaxed and more playful approach of technology than women [5]. Other studies have detected differences between boys and girls already in the child age. A study of Chen [8] shows that boys perceive computer science to be more useful than girls and thus show more motivation as a result. Girls, however, often better understand the concepts and perform better than boys.

Based on his empirical investigations, J.M. Keller developed the *ARCS Model of Instructional Design* as a "method for improving the motivational appeal of instructional materials" [18]. The model contains four conceptual categories that "subsume many of the specific concepts and variables that characterize human motivation" [18]: *Attention, Relevance, Confidence* and *Satisfaction*. Additionally, it proposes strategies to enhance the motivational appeal of instruction. In summary, it sketches a systematic design process, called *motivational design*. The four conceptual categories are defined by Keller as follows [18]:

- *Attention* has not only to be directed to the appropriate stimuli, but also sustained during the learning process.
- *Relevance* provides the answer to the question "'Why do I have to study this?"
- *Confidence* can influence a student's persistence and accomplishment. Confident people tend to believe that they can effectively accomplish their goals by means of their actions, while unconfident people want to impress others and worry about failing.
- *Satisfaction* makes people feel good about their accomplishments.

3   The Bebras Challenge

The first Bebras Challenge was organized in October 2004 in Lithuania by V. Dagiene [11]. The goal of Bebras "is to promote Informatics (or Computer Science, or Computing) and Computational Thinking especially among teachers and pupils of all ages, but also to the public at large by extent"[1]. Up to now, the project has reached a tremendous dissemination, attracting more than 1 300 000 participants from 35 countries in 2015[1]. Bebras is offered annually in these states, separated in different age groups. Although the test is performed usually in schools and the students have to be registered by their teachers, the participation is voluntary [4]. The students are free to work on their own or in a team. Each age group has to solve a different set of tasks, which are created by the national Bebras boards, which also classify them in one of the difficulty levels *easy, medium* or *hard* [9]. Some tasks are offered to more than one age group, providing opportunities to compare these age groups at least in some regard. For an actual, comprehensive overview see [10].

---

[1] www.bebras.org.

The German challenge is performed annually during one week in November in all 16 federal states, separated in 4 age groups for students of grades 5/6, 7/8, 9/10 and 10–13 respectively. During this week, students can solve the 18 tasks of their age group on any day. A user or a team has to solve all tasks during 40 min from logging in on the website. When starting, each participant or team has a credit of 54 points. For solving a task correctly, depending from its difficulty level, 6, 9, or 12 points are added. For wrong solutions, 2, 3 or 4 points are subtracted. At the end, each participant or team can achieve a total score from 0 to 216 points.

4   The Data

In this study, we have analyzed the results of the of the German Bebras challenge of 2014 with 217 604 registered users, see Tables 1 and 2[2].

Table 1. Age group distribution

	Grade 5, 6	Grade 7, 8	Grade 9, 10	Grade 11–13	Total
Number	52 465	71 331	62 915	30 795	217 506
Percentage	24.1 %	32.8 %	28.9 %	14.2 %	100 %

Table 2. Gender distribution over the age groups

	Grade 5, 6	Grade 7, 8	Grade 9, 10	Grade 11–13	Total
Boys	51.1 %	53.0 %	58.8 %	67.7 %	56.3 %
Girls	48.8 %	46.7 %	40.8 %	32.2 %	43.4 %
Not ind	0.2 %	0.3 %	0.4 %	0.2 %	0.3 %

Basically, the responses of the individual participants to the tasks are assigned to teams. Every team consists of one or two members. We restricted our analysis to teams that had responded at least to one task. By this restriction, the number of analyzed participants was reduced to 214 811.

Regarding the team size, 60.7 % of the participants had worked alone, while 39.3 % had preferred to collaborate in pairs. As expected, the percentage of collaborating girls (39.2 % to 41.1 % over the age groups) was higher compared to the boys (from 35.0 % to 37.7 %). According to a χ^2-test on the fourfold tables *gender × team size*, the hypothesis that *team size* and *gender* were independent could be rejected in all age groups with probabilities below 0.01.

For the analysis of the individual performances, we produced a dichotomous matrix for each of the 4 age groups, consisting of one line per participating team (one or two persons) and one column per task of the respective age group,

---

[2] www.informatik-biber.de/archiv/informatik-biber-2014.

18 in total. In the cells, we entered 1 if the respecting team solved this task correctly or 0 in all other cases (incorrect answer, no response, or task not even opened).

To calculate the empirical difficulty of every task, we calculated the average over each column of this matrix, which is equivalent to the relative frequency of correct responses to the respective task. For the average success rates of any subset of participants (girls or boys, singles or pairs) in a certain task, we calculated the average of the respective column restricted to this subset, e.g. by filtering all single girls among the participating teams.

We restricted this analysis to students that have indicated their gender, working alone or in same-gender teams, excluding mixed-gender groups by this way. Over all age groups, the portion of mixed pairs among all pairs increased from 5.2 % in the youngest to 14.3 % in the eldest age group.

5   Results

5.1   Differences over All Tasks

Our first goal was to compare the performances of boys vs. girls and of same-gender pairs vs. singles over all tasks. For this purpose, we could have used the Bebras overall score. Yet, as explained above, the Bebras score values depend from the difficulty levels of the tasks, as classified by the Bebras boards. The calculation of the Spearman Rank Correlation [6] between the empirical difficulty (average over all participants on our 0/1 scale) and the task level as classified by the Bebras board for all four age groups (*easy*, *medium*, *hard*) varied from 0.41 to 0.81 over the four age groups. Thus we did not regard the Bebras overall scores as sufficiently valid for the individual performance. Instead, we calculated the average performances directly from our dichotomous matrices, without any weighting of the tasks. For this purpose, we subtracted the averages over the compared subgroups, e.g. the average over all single girls from the average over all single boys, see Table 3.

To test the significance of the detected differences according to the ordinal level of measurement (19 discrete values between 0 and 1), we applied a two-sided Wilcoxon Rank-Sum test, see [25]. For the rejection of the hypothesis that the distributions of two subgroups were equal, we choose a significance level of of $p < 0.05$. According to this level, all the differences indicated in Table 3 turned out to be significant.

As the results in Table 3 show, the boys performed significantly better in all age groups compared to the girls, as well working alone as in pairs. Yet, the differences increase dramatically from the youngest up to the eldest group. For both gender groups, the pairs were more successful than the single participants. Surprisingly, the difference between paired and single girls was lower compared to the difference between paired and single boys. In this regard, we could not confirm the results of [7] and [14].

Table 3. Significant differences of average performance over all tasks ($p < 0.05$)

	Grade 5–6	Grade 7–8	Grade 9–10	Grade 11–13
SingleBoys-SingleGirls	0.004	0.021	0.055	0.086
PairedBoys-PairedGirls	0.008	0.037	0.057	0.102
PairedBoys-SingleBoys	0.070	0.074	0.067	0.064
PairedGirls-SingleGirls	0.066	0.059	0.065	0.049

5.2 Differences on Task Level

To analyze the gender differences on task level, we subtracted the average performance of the girls from the average performance of the boys for each task. This was performed separately for single participants and pairs in each age group. In the two elder age groups, the boys had outperformed the girls in all 18 tasks. Yet, in the two younger age groups, the girls worked better in several tasks. Therefore, we restricted our analysis to these age groups. For significance, we applied a two-sided Wilcoxon-Rank-Sum test again (see Sect. 5.1). Because two tasks (295 and 335) had differences that exceeded the usual significance level $p = 0.05$ very closely, we raised this level slightly to $p < 0.057$.

Altogether, 27 different tasks were presented to the two younger age groups, of which four (307, 334, 337, and 340) did not show any significant difference between the performance of girls and boys. In grades 5 and 6, the differences were nearly symmetric: the girls outperformed the boys significantly in seven tasks, as singles as well as in pairs. In turn, the boys were significantly better in eight tasks, also independent of team size. The girls of grades 7 and 8, showed a significantly better performance compared to the boys in 3 tasks, in one of these (288) only as singles. Two tasks (303 and 308) seemed to be particularly attractive to girls, because they were solved significantly better by single as well as by paired girls in both inspected age groups. The boys of grade 7 and 8 outperformed the girls in 13 tasks significantly, in one of these (309) only as singles, in another (316) one only in pairs.

For the following analysis, we grouped the 27 tasks as follows, see Table 4:

1. *Girls Tasks*: all tasks that were solved significantly better by single *and* paired girls,
2. *Boys Tasks*: all tasks that were solved significantly better by single *and* paired boys,
3. *Neutral Tasks*: the rest of the tasks, either without any significant gender difference or showing such a difference in only one case (singles *or* pairs), containing the tasks 288, 307, 309, 316, 334, 337, and 340.

Table 4. Absolute values of significant differences in the performances of girls and boys for the groups *Girls Tasks* and *Boys Tasks* ("NA" means "task was not available in this age group")

TaskNr	Grade 5–6 Singles	Pairs	Grade 7–8 Singles	Pairs
Girls tasks				
294	0.033	0.030	NA	NA
295	0.010	0.019	NA	NA
296	0.020	0.020	NA	NA
300	0.044	0.033	NA	NA
303	0.022	0.012	0.019	0.008
308	0.033	0.035	0.024	0.020
327	0.060	0.050	NA	NA
Boys tasks				
292	NA	NA	0.048	0.075
297	0.088	0.085	NA	NA
299	0.031	0.036	NA	NA
301	0.013	0.028	0.025	0.045
302	0.037	0.036	0.038	0.031
304	0.024	0.039	0.049	0.057
311	NA	NA	0.009	0.018
313	NA	NA	0.025	0.053
323	NA	NA	0.083	0.114
333	0.034	0.052	0.058	0.073
335	0.011	0.014	0.030	0.050
336	0.054	0.059	0.082	0.098
338	NA	NA	0.009	0.028

6   Application of the ARCS Model

As explained above, most participants were encouraged to participate in Bebras by their teachers. Yet, the students solved the tasks individually on the computer without direct control of their teachers. Therefore, the individual motivation of the students might play a dominant role for their performance. To investigate this role, we will analyze the tasks[3] according to the motivation factors of Kellers ARCS Model [17], see Sect. 2. As the effects of *Satisfaction* seemed relevant predominantly regarding subsequent challenges, we will drop this factor here.

---

[3] www.informatik-biber.de/archiv/informatik-biber-2014.

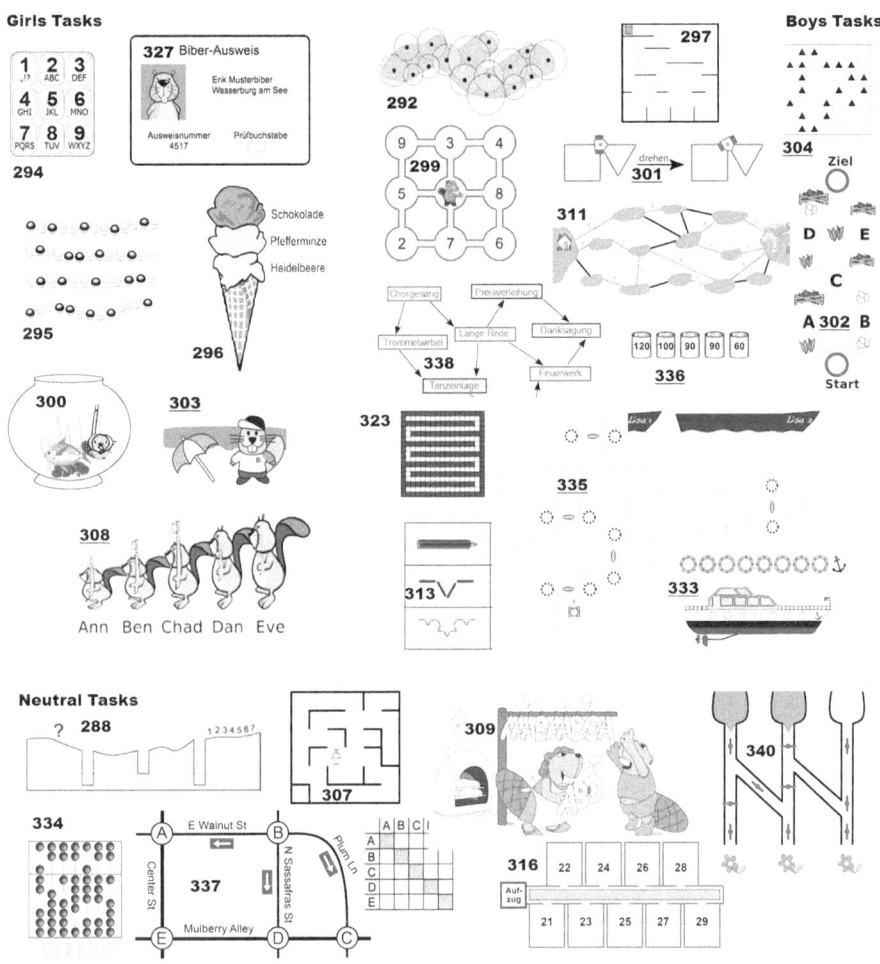

Fig. 1. Pictures of the tasks

6.1 Attention

According to the ARCS model [17], *attention* is a relevant factor for motivation and, in consequence, also for success. Assuming that attention has to be attracted by each task for itself, its first glance appearance is likely to be crucial. Therefore, most of all, its graphical elements like pictures or diagrams, will be relevant for the attention it gets (see Fig. 1). Several task numbers are underlined to demonstrate that these tasks had produced differences in all four columns of Table 4.

Looking at the graphical elements of the Girls Tasks, we find that these are mostly representing animals, jewelry or food. The only exceptions are task 294 (keyboard) and partly 327, which displays a beaver, but also the rest of an id card. It is apparent that most of these pictures look at least partly lovely and likable, except task 294. Yet, the latter might be attractive for girls by the relevance of the picture, assuming that girls like to use their phones (see below). Regarding the Boys Tasks, the dominating elements are mostly abstract rectangular figures, graphs or technical apparel. The Neutral Tasks have an appearance that is more or less similar to the Boys Tasks, except task 309, which looks more like a girls task. Yet, this task is relatively difficult, which may have caused a low confidence level of the girls, see Sect. 6.3.

6.2   Relevance

According to The Glossary of Education Reform [1], "in education, the term relevance typically refers to learning experiences that are either directly applicable to the personal aspirations, interests, or cultural experiences of students (personal relevance) or that are connected in some way to real-world issues, problems, and contexts (life relevance)". Applied to the Bebras tasks and the personal situation of 10–13 year old students, these criteria might be represented by the closeness of the situation that is described in the Bebras task to the personal life and experience of the students. In other words, the relevance will be determined by the probability that the participants or their friends have experienced or will experience a similar situation. Keller [17] describes this by the term *Familiarity*.

In this sense, obviously, all of the Girls Tasks have a certain relevance, referring to situations, which actually could occur to the children in their daily life:

- 294: How to write the name of a friend on a phone keyboard?
- 295: How to identify your bracelet?
- 296: How to order scoops of ice cream according your preference?
- 300: How to assemble stickers to the picture of an aquarium?
- 303: How to find out which photograph your friend wants to get?
- 308: How can kids get the proper toothbrush?
- 327: How to secure your id card?

On the other hand, all of the Boys Tasks lack this relevance based on everyday experiences in some regard, at least for girls:

- 292: How to cover a territory by mobile phone transmitting masts?
- 297: How can a robot cross a labyrinth?
- 299: How many friends can a Beaver visit in four days in a given rectangular (!) set of channels and ponds?
- 301: How can a draw bot draw a certain figure?
- 302: Which path can a Beaver travel in a polygonal river system with a certain amount of Energy?
- 304: How many mobile phone transmitting masts have to be installed to cover all houses of a certain village?
- 311: How to find the cheapest way over a set of toll bridges?
- 313: How to generate trunk patterns?
- 323: How to catch a monster in a labyrinth?

- 333: How to arrange portholes of different colors?
- 335: How to move a robot on a certain path?
- 336: How to optimize the payload of a ship?
- 338: How to synchronize the events of an abstract ceremony?

The Neutral Tasks 307, 334 and 337 lack this relevance more or less, in contrary to the tasks 288, 309, 316 and 340. Yet, 288, 316 and 340 might fail to attract girls by their comparably abstract pictures, while 309 is quite difficult and thus could reduce the confidence level (see below).

- 288: How can Beavers cross potholes?
- 307: How to move a robot to a certain point
- 309: How many pretzels have been sold?
- 316: How to optimize walking distance to hotel rooms?
- 334: Which number is represented by this constellation?
- 337: How to represent one- and two-way streets in a matrix?
- 340: Which flower gets water in this constellation of valves?

According to Kellers' Model, for both genders relevance should support motivation. Yet, as shown by [23], boys tend to like problem solving for itself much more than girls, see Sect. 2, which could explain the better performance of the boys in the comparably irrelevant tasks.

6.3 Confidence

One of the relevant factors of confidence is the likelihood of success with a given amount of effort and ability [17]. In consequence, the apparent difficulty of a task will influence the motivation to solve it. To assess the difficulty of a task before solving it, the participants can take this information directly from the classified level (as displayed by the Bebras system) or guess how difficult the solution might be. Therefore, we analyzed the empirical task difficulty (in other words the average performance over *all* participants) as well as the difficulty level classified by the Bebras board. As already mentioned in Sect. 5.1, the Spearman Rank Correlation [6] between these two values varies strongly (0.76 in the youngest age group, but only 0.41 in the group of grades 7,8). Table 5 compares these two difficulty measures for the Boys and the Girls Tasks. Please note that a task is the more difficult so solve, the lower the values of empirical difficulty (solution

Table 5. Comparison of task difficulties

	Grades 5–6 Boys tasks	Girls tasks	Grades 7–8 Boys tasks	Girls tasks
Overall empirical difficulty	0.35	0.82	0.37	0.91
Classified easy	13 %	57 %	27 %	100 %
Classified medium	25 %	43 %	36 %	0 %
Classified hard	63 %	0 %	36 %	0 %

Table 6. Empirical and classified difficulties of the neutral tasks

TaskNr	Grades 5–6 Empirical diff	Classified	Grades 7–8 Empirical diff	Classified
288	NA	NA	0.58	Hard
307	NA	NA	0,51	Medium
309	NA	NA	0.47	Hard
316	NA	NA	0,49	Medium
334	0,52	Medium	0,67	Easy
337	0,09	Hard	NA	NA
340	0,83	Easy	NA	NA

percentage) are. Apparently, the girls tend to perform better in easy and medium tasks. As an explanation, the self-efficacy of girls was found to be lower compared to boys [19], see Sect. 2, therefore the girls might drop difficult tasks. On the other hand, the boys show higher willingness to deal with challenging problem solving activities by trial and error [16], see Sect. 2.

Table 6 displays the difficulties of the Neutral Tasks, which might explain, why both genders are comparably successful. Task 288 looks like a typical boys task, supported by its difficulty. Yet, its high relevance could have motivated the girls also. The tasks 307, 316 and 334 have medium difficulty and a quite abstract or technical appearance. While 316 is comparably relevant, 307 and 334 are not. The task 309 looks like a girls task, but due to its difficulty, the confidence of the girls might have been low. In case of task 337, the extraordinary difficulty might explain the missing differences between boys and girls, as only 9 % of all participants have solved this task at all. The remaining task 340 seems to attract both genders equally, by combining easiness, a "male" picture and a situation comparably relevant for girls.

7   Conclusion and Future Work

In this paper, we have demonstrated that three factors of motivation according to Kellers ARCS-Model are particularly relevant for the performance of the 10–13 year old girls. This result could be utilized by the Bebras boards to construct tasks that motivate particularly the younger girls: (1) look for a situation that is likely to occur in girls' everyday life, (2) construct a task for this situation that is not too difficult and (3) draw a nice picture that contains a person, an animal or other lovely objects. This might be also a good advice for computer science teachers looking for tasks that are motivating for younger girls.

Nevertheless, our study has some weaknesses. Most important, we did not decide between unanswered or incorrectly answered tasks, although the psychological reasons might be very different in both cases. Further studies could analyze this effect as well as apply our methodology to other Bebras Challenges, as there are about 34 each year around the world.

References

1. Abbott, S. (ed.): The glossary of education reform. http://edglossary.org/relevance/
2. Bandura, A.: Self-efficacy: toward a unifying theory of behavioral change. Psychol. Rev. **84**(2), 191–215 (1977)
3. Bandura, A.: Self-efficacy. In: Encyclopedia of human behavior, vol. 4, pp. 71–81. Academic Press, New York (1994)
4. Bebras Community: Statutes - RC3 (2015). http://www.bebras.org/sites/default/files/BebrasStatutes_rc3.pdf
5. Beyer, S., Rynes, K., Perrault, J., Hay, K., Haller, S.: Gender differences in computer science students. In: Grissom, S., Knox, D., Joyce, D., Dann, W. (eds.) Proceedings of the Thirty-Fourth SIGCSE Technical Symposium on Computer Science Education. SIGCSE bulletin, vol. 35(1), pp. 49–53. ACM, New York (2003)
6. Spearman, C.: The proof and measurement of association between two things. Am. J. Psychol. **15**(1), 72–101 (1904). http://www.jstor.org/stable/1412159
7. Cen, L., Ruta, D., Powell, L., Ng, J.: Does gender matter for collaborativelearning? In: International Conference on Teaching, Assessment and Learning (TALE), pp. 433–440. IEEE, Piscataway (2014)
8. Chen, M.P.: The effects of prior computer experience and gender on high school students' learning of computer science concepts from instructional simulations. In: Jemni, M., Kinshuk, S.D., Spector, M.J. (eds.) Proceedings 10th IEEE International Conference on Advanced Learning Technologies, pp. 610–612. IEEE, Los Alamitos (2010)
9. Dagiene, V., Futschek, G.: Bebras international contest on informatics and computer literacy: criteria for good tasks. In: Mittermeir, R.T., Syslo, M.M. (eds.) Informatics Education - Supporting Computational Thinking. LNCS, vol. 5090, pp. 19–30. Springer, Berlin (2008)
10. Dagiene, V., Stupuriene, G.: Bebras - a sustainable community building model for the concept based learning of informatics and computational thinking. Inform. Educ. **15**(1), 25–44 (2016)
11. Dagiene, V., Žalys, D.: Bebras - informaciniu technologiju konkursas. Kompiuterija **11**(87), 50 (2004)
12. Deci, E.L., Â Ryan Richard M.: Intrinsic motivation and self-determination in human behavior. Plenum Press, New York (1985)
13. Graham, S., Latulipe, C.: CS girls rock: Sparking Interest in Computer Science and Debunking the Stereotypes. In: Grissom, S., Knox, D., Joyce, D., Dann, W. (eds.) Proceedings of the Thirty-Fourth SIGCSE Technical Symposium on Computer Science Education. SIGCSE bulletin, vol. 35, pp. 322–326. ACM, New York, N.Y. (2003)
14. Hubwieser, P., Muhling, A.: Investigating the Psychometric Structure of Bebras Contest: Towards mesuring Computational Thinking skills. In: Lee, G.C., Berglund, A., Wuh, C.C. (eds.) Proceedings of the 2015 International Conference on Learning and Teaching in Computing and Engineering, pp. 62–69. IEEE, Los Alamitos, California (2015)
15. Joshi, K.D., Schmidt, N.L.: Is the information systems profession gendered? characterization of IS professionals and IS career. ACM SIGMIS Database **37**(4), 26–41 (2006)
16. Katz, S., Aronis, J., Allbritton, D., Wilson, C., Soffa, M.L.: Gender and Race in Predicting Achievement in Computer Science. IEEE Technology and Society Magazine **22**(3), 20–27 (2003)

17. Keller, J.M.: Motivational design of instruction. In: Reigeluth, C.M. (ed.) Instructional-Design Theories and Models: An Overview of Their Current Status, Erlbaum, Hillsdale, NJ [u.a.], vol. 1, pp. 383–434 (1983)
18. Keller, J.M.: Development and use of the arcs model of instructional design. Journal of Instructional Development **10**(3), 2–10 (1987)
19. Lamoureux, I., Beheshti, J., Abuhimed, D., AlGhamdi, M.J.: Gender Differences in Inquiry-Based Learning at the Middle School Level. In: Grove, A. (ed.) Proceedings of the Annual Meeting of the Association for Information Science and Technology. Association for Information Science and Technology, Maryland (2013)
20. Madigan, E.M., Goodfellow, M., Stone, J.A.: Gender, Perceptions, and Reality: Technological LiteracyAmong First-Year Students. In: Russell, I., Haller, S., Dougherty, J.D., Rodger, S. (eds.) Proceedings of the 38th SIGCSE technical symposium on Computer science education, pp. 410–414. ACM, New York, NY (2007)
21. Murphy, L., Richards, B., McCauley, R., Morrison, B.B., Westbrook, S., Fossum, T.: Women catch up: Gender Differences in LearningProgramming Concepts. In: Baldwin, D., Tymann, P., Haller, S., Russell, I. (eds.) Proceedings of the thirty-seventh SIGCSE Technical Symposium on Computer Science Education, pp. 17–21. ACM Press, New York (2006)
22. Patitsas, E., Craig, M., Easterbrook, S.: A historical examination of the social factors affecting female participation in computing. In: Cajander, Å., Daniels, M., Clear, T., Pears, A. (eds.) Proceedings of the 2014 conference on Innovation and technology in computer science education, pp. 111–116. ACM, New York, N.Y. (2014)
23. Redmond, K., Evans, S., Sahami, M.: A large-scale quantitative study of women in computer science at Stanford University. In: Camp, T., Tymann, P., Dougherty, J.D., Nagel, K. (eds.) Proceeding of the 44th ACM Technical Symposium on Computer Science Education, pp. 439–444. ACM Press, New York (2013)
24. Schelhowe, H.: Gender questions and computing science. In: Morrell, C., Sanders, J. (eds.) Proceedings of the international symposium on Women and ICT creating global transformation. ACM Press, New York (2005)
25. Wilcoxon, F.: Probability tables for individual comparisons by ranking methods. Biometrics **3**, 119–122 (1947)

Attitudes Towards Computer Science in Secondary Education: Evaluation of an Introductory Course

Daniel Lessner[✉]

Charles University, Prague, Czechia
`lessner@ksvi.mff.cuni.cz`

Abstract. Computer science (CS) is not being taught at Czech grammar schools (15–18 years old students). In our effort to change that, we developed and piloted a basic CS course. It introduces the fundamental ideas of CS comprehensively and in relation to existing subjects and real world applications. In this paper we describe the course program briefly and present the part of evaluation that focuses on students' point of view. We assessed how they perceive the subject of CS in context of other school subjects using a questionnaire with a qualitative and a quantitative part.

It turns out that our approach to CS is considered intellectually very demanding, yet this does not seem to affect other features (interest, usefulness) too negatively. CS does not show extreme values in other measured attributes in comparison to other science subjects.

Keywords: Qualitative study · Computer science education · Secondary education · Student attitudes

1 Introduction

Computer science (CS) is not included in general education in Czechia. Even though we have subjects with "informatics" in their name, they focus mostly on developing digital skills. We will refer to these subject as "ICT" for the sake of clarity. The commonly shared idea is that CS is a specialized field, irrelevant and too complex for everyday life. One of the results of this is that the students are not even aware of the existence of CS as a study of efficient information processing.

In order to improve this situation, we developed a curriculum for introductory CS which would fit our grammar schools (15–18 years old students). The general approach towards our project stemmed from design based research. We have iteratively tested, evaluated and improved the CS programme. It gave us empirical ground to answer questions such as "What can CS look like at Czech grammar schools?", "Can the students even manage it?" and "What good can that bring?".

© Springer International Publishing AG 2016
A. Brodnik and F. Tort (Eds.): ISSEP 2016, LNCS 9973, pp. 53–64, 2016.
DOI: 10.1007/978-3-319-46747-4_5

In this paper we examine how students perceive the subject and why. We wanted to further investigate the aspect of difficulty, as it was coming up repeatedly during the teaching in conversations with both students and colleagues. We also wanted to know whether students noticed improvements in problem solving and communication skills and how do they regard CS in comparison with other subjects. However, as the nature of our survey was partially qualitative, unexpected topics emerged as well.

The paper begins with a brief review of relevant work. Then we describe the conditions of our experimental teaching. Section 4 deals with the process of collecting data for our research and Sect. 5 describes the process of evaluation. Section 6 introduces the results logically organized in topical subsections. A short section of recommendations for similar surveys is the last before final conclusions.

2   Relevant Previous Work

Here we summarize what is known about attitudes towards rigorous CS at schools and how does our work relate to it. We proceed from general studies to those more specifically related to this paper. A general overview of research on students' attitudes towards scientific school subjects is given in [11]. Klopfer in [7] categorizes affective behaviors in science education and distinguishes attitudes towards science and scientists, acceptance of scientific methods, enjoyment of learning science and developing interest in science related activities and careers. It needs to be pointed out that "scientific attitudes", "attitudes towards science" and "attitudes towards school science" are three related, but different constructs. We are interested in the latter in this paper.

Furthermore, attitude is not the only nor the dominant determining factor of behavior [11]. The context of the situation matters. To link attitudes and behavior, stronger understanding of attitudes and the context is necessary. This implies the need to employ qualitative methodologies for exploring specific issues of students' attitudes to school science [11,12]. We took their recommendations into account in our study.

A study of 600 students compares different groups to find out that their expectations depend on prior experience with computers and that they are gender specific [14]. In a detailed quantitative evaluation of 238 students attitudes towards programming, Czech and Slovak students assessed the subject of programming as "rather interesting", "rather necessary" and "rather usable" for their future [19]. However, they are students of schools with special focus on programming, ICT and CS.

One of the few large and complex studies in our country is [13]. The authors conclude that teachers consider as least important those topics which they also understand the least, such as algorithms and programming, effectively developing and supporting a misplaced view of the subject among pupils. The authors also tried to examine pupil attitudes to specific topics. Many topics appear to be alternately both liked and disliked. Pupils seem to consider topics which include systematic work and use of memory and logic as less important, not fun, and difficult. This includes virtually anything beyond digital skills.

A different source [10] is based on data collected after the Bebras contest [4]. Once again, answers to open questions clearly show that many pupils and students hold a rather misguided conception of CS: according to them, Bebras is not even a CS contest, because there is too much logic and thinking, and too few computers.

The closest study to ours is [17]. Its authors investigate changes in students' views following CS lessons based on the CS Unplugged activities [2]. It works with a small sample of students of similar age. Moreover, many (not all) of the activities we use are unplugged and adhere to similar principles as the CS Unplugged resources. The study found that even though the activities improved students' understanding of the nature of CS, the intentions to study it further actually declined.

3   Preconditions, Decisions

In this section we describe the circumstances under which our experiment took place, loosely following the Darmstadt Model [6] where appropriate.

3.1   Organizational Aspects

Grammar schools, also known as "Gymnasium" in Central Europe and Germany, are one of the branches of secondary education in Czechia. They focus mostly on traditional academic disciplines and should prepare the student to continue at virtually any type of university. Their programme is defined in the Framework Educational Programme (FEP) [1]. Mentions of proper CS in FEP are sparse and confused and thus usually not reflected in the school programmes. This was the case on our pilot school as well.

One of the compulsory subjects is called "informatics" (ICT). Although in reality, students learn there how to use digital technologies. Our class had one and two lessons of ICT per week in the two previous years, respectively. In our school year we had one lesson (45 min) each week. To fit in the classroom with computers, the class was split into halves which had their lessons separately.

The course was taught by the author. This was a necessity in the given organizational constraints, not a deliberate choice. To reduce the negative effect on the quality of our research, we took detailed notes immediately after the lesson (or during the lesson, when possible), sometimes recording it, anonymizing the questionnaire data and discussed the methods and results with other researchers.

3.2   Sociocultural Related Factors

The common conception of computer related education in both primary and secondary education revolves around consuming the technology. "Actual" or "proper" CS is generally interpreted as coding, what is something unattainably abstract and overall useless for anyone but programmers. This applies also for our pilot school, where our colleagues perceived our task as "mainly teaching them all those applications". The subject is regarded as auxiliary.

Our 32 students were 15–16 years old, 17 females and 15 males. They were in the first year of the second half of the 8-year cycle, i.e. in the first year in actual secondary education. Our lessons were compulsory. This ensured a full range of students' characteristics. A few students had some previous experience with programming, some were in advance already strongly uninterested in CS.

3.3   Educational Objectives and Content

Here we describe briefly the course plan itself. The objectives are derived from the general objectives for Czech grammar schools: developing key competences (the most relevant for CS are problem solving, communication and learning) and providing general knowledge. The course should introduce CS to grammar school students in a similar manner as other scientific subjects do, as promoted for example in [5]. Students should know the basic terms, topics, and questions of the field, methods it uses, fundamental results, open problems, and applications and connections to other areas. They should be able to apply their knowledge and skills efficiently also when solving ordinary problems in everyday life (outside CS), exhibiting basic computational thinking [18].

To achieve such goals, we have found appropriate fundamental ideas in the spirit of [3,15]. It seemed more efficient than adapting existing curricula into our constraints (even though we worked with them as well). The course covers the following topics on an introductory level:

1. Information: what it is, efficient questions, measuring, encoding, binary system, decision trees
2. Graphs: what are they good for, representation, Eulerian paths, isomorphism
3. Problem solving and state space: representing a problem, traversal, limitations
4. Algorithms: properties, representations (free text, flowcharts), distinguishing algorithms, existence of limitations
5. Programming: automating calculations with Python, basic flow control structures
6. Efficiency: counting steps, upper bounds, sorting examples and efficient processing of sorted lists, recognizing and avoiding exponential processes
7. Advanced topics: topological sorting and critical path method, Turing test and the current state of artificial intelligence, expected development

Presented topics may seem overly complex for the given age and time frame. We indeed had to simplify them significantly. The most abstract topics (e.g. halting problem) is regarded as similarly abstract topics in other subjects (e.g. modern physics): students should be aware of it and know some implications, but we do not expect everyone to understand it fully. More details on the course are provided in [8]. We also develop an online textbook[1] which will give a virtually complete specification [9].

---

[1] With the kind support of Google's CS4HS program.

4   Data Collection

In this section we describe the process of collecting the data. Some comments on the actual teaching are necessary first, since it somewhat differed from the plan. The timetable only allowed 45 min of weekly direct teaching instead of the expected 90. We had to prioritize our goals and add more homework than planned, sacrificing popularity for mastery. There was some homework almost every week, and it was designed to last approximately 30–45 minutes. Even though we have discussed it with the students beforehand, they were surprised and found the homework to be a burden.

One of the measures we took to help students with the transition from ICT to CS was a questionnaire and focused discussion at mid-term. The aims were to investigate the students' difficulties and to find some solutions and recommendations. We used the results to better inform the design of this study as well. One of the unintended positive side-effects was that based on their previous experience, students considered the final questionnaire as something potentially useful and worthwhile. They also knew that open and honest answers do not pose any threat for them.

At the very end of the school year we organized an extra lesson to give out the questionnaire we designed for this purpose. We checked whether students understood the questions correctly and extended the instructions where necessary to increase the validity of their answers. The grades were already determined and the students knew it (so we could expect credible and authentic answers). We collected 31 questionnaires from the total of 32 students.

The questionnaire had a qualitative and a quantitative part and consisted of both open and closed questions. Literature (such as [11]) shows how misleading can quantitative studies be. Further explanations and reasoning was encouraged everywhere. It covered mainly the following areas:

- What is CS
- Emotions related to CS lessons
- Key competences developed during CS lessons
- Attitudes towards specific topics
- Thorough comparison of CS and other subjects

The areas, comparison criteria and specific questions were based on previous interviews with students, the experiences from mid-term and partially on the categorization given in [11,12]. The questionnaire gave us certainty that every voice is heard (unlike in a group discussion) while keeping a reasonable time frame within the regular school schedule (unlike individual interviews). We learned what lays in the background of students' answers and also how frequent are individual attitudes (because informal discussions often give a rather distorted view).

5   Data Processing and Evaluation

Our approach to analyze the answers from open questions was based on open coding [16], a technique used also in [13]. Meaningful fragments of text are tagged with a code, guided by our interests (e.g. motivations, challenges, attitudes etc.). The code serves as a shorthand for the meaning of the text fragment. Each code represents a group of fragments. Our data consist of relatively short texts structured according to the questionnaire, what makes the process of coding simpler. The disadvantage of brief answers is that we can not get very deep in our interpretations.

The process of coding is done iteratively, the texts are reread and recoded. Codes with similar meanings are joined, new codes are discovered. The meaning of the codes can shift and develop during the process. To improve the research validity, we recoded the texts multiple times, until the converged meaning was used consistently. The resulting codes give an overview on what do the texts talk about.

It is often argued in Czechia that CS is something just too extreme to be included into mainstream education. We were therefore very interested where would students put it relative to other subjects in terms of (self-reported) popularity, interest, usefulness, difficulty of subject matter, difficulty of achieving a good grade and level of own achieved mastery. As the rest of the questionnaire, these attributes were based mainly on our previous interviews with students and findings of the mid-term survey [8]. This constituted a table with these attributes as rows and compared subjects as columns. An extra row was present for similarity of all the subjects to CS, and an extra column for ICT from the previous year. Students were filling in values of 1–100 to the cells. Our reason to prefer scores over more usual rankings is the possibility to represent various distributions of each attribute with such system and obtain more meaningful data (as suggested in [11]).

The reported values allowed us to quantify the attributes in context with other subjects. We have also compared groups of subjects, such as languages, humanities and sciences. We used averages, quartiles, extremes and rank measures for comparisons. We will be discussing medians unless we say otherwise, because the values are most often nicely distributed so even medians describe the measured aspects sufficiently. With a sample of our size and character any more advanced statistics would be meaningless.

Aside from the questionnaires, we worked with our notes taken immediately after lessons, with data in Moodle (e.g. submitted assignments, achieved points) and with the school agenda (final grades from all subjects, absences). These extra sources together with some redundancy in the questionnaire (e.g. asking very related questions) allowed us to confirm that students' opinions and other data stand in accordance and thus helped the reliability of our results. They also added more context and allowed deeper understanding of the answers.

6   Results

The information obtained from all the available data is too much to describe completely in this paper, so we describe only the most interesting findings. We organized this section along topics which arose from the evaluation in the spirit of qualitative methodology. The rate of achieving educational goals is not the topic of this paper, yet a brief comment is important for interpretation. There is of course room for improvement, but the students have accomplished the course goals satisfactorily overall. This claim is based on examining their grades, in-class performance and assignments. They did not master 100 % of the matter, but their skills and understanding fall within the range of other subjects

6.1   Difficulty, Emotions, Homework and Popularity

Let us begin with the deepest of the examined aspects. Students answered the question "What inspires your strongest emotions when studying informatics, end which emotions are they?" (note that "informatics" is the official name of the subject, both for our CS course and the previous ICT). A feedback loop known from our small mid-term research emerged: failure inspires negative emotions, success (solving a problem, understanding a concept) inspires positive emotions. However, a closer look revealed more details about the sources of those feelings.

Negative emotions reports outnumbered the positives. Students mention despair, fear, anger, sadness and others, mostly when facing homework. The issue of homework came up also in explanations of popularity. Apparently, homework was a high ranking factor in decreasing reported popularity, even though only a minority of students saw homework as an issue.

Students reported also more specific factors than the mere existence of homework: incomprehension of instructions, existence of deadlines, amount of work and subject difficulty were among the sources of their issues. Inability to understand the instructions was reported most frequently. But when investigated individually, students actually were able to read and understand *what the question is* sufficiently. The true source of their confusion was that we did not tell them explicitly and exactly *what to do*.

Almost all students say that once it was clear what to do, the homework was doable, the matter understandable and interesting and a good grade achievable. This is in accord with mastery both measured in grades and declared in the questionnaire. Students' answers suggest that with better balanced difficulty and workload and more carefully formulated tasks, their overall experience would significantly improve.

Other often declared factors determining reported popularity are usefulness of CS and interest towards CS. This is supported also by students' reasoning about specific topics and by higher mutual correlations of these attributes in reported quantitative data (both ca. 0.65). All these attribute correlations are stronger in CS then in average (over all subjects). While popularity, usefulness, interest and difficulty clearly are related (all our data suggest that), none of them completely determines any other. We can not rely on e.g. emphasizing usefulness to increase interest in all students.

6.2 Key Competences

We let students to asses the change in their ability to solve problems, to communicate efficiently and to study. Students were to make a mark in the form on a scale from *decrease* over *no change, slight increase* and *significant increase* to *maximum possible increase*. This asymmetry is based on our previous knowledge: virtually no one perceives a decrease, the question is how big is the perceived increase.

As a whole (considering mean, median and mode), students declare *slight increase* in problem solving skills. The other two abilities also changed positively, but the effect is not considered that strong. No one declared any decrease. The improvement in communication (where declared) is attributed mostly to writing algorithms. The improvement in studying is attributed rather to the process than to the subject itself: the necessity to actually work, to meet the deadlines etc.

Most interesting answers are related to problem solving. We asked about the cause of the changes. Students say they learned new approaches to solve problems, new useful technologies (seen as a tool, not the subject). Many mention efficiency as a criterion to judge different approaches. Some say training and examples, some say motivation and fun doing it. However, most express in one way or another that improvement in problem solving is an inherent feature of studying computer science.

6.3 CS, ICT and Other Subjects

Until we started our experimental teaching, the subject named informatics focused on using ICT. So we were curious to see how does the previous approach compare with ours in the eyes of our students. Changes were organizational (home assignments, employment of Moodle), methodological (complex tasks requiring independent decisions instead of following given instructions) and of course in content. Students suddenly had to think on a different level and deal with a new kind of problems. Technology was less of a subject to study and more of a tool to use. Efficiency became a fundamental issue.

Here we briefly comment on the quantitative data. We discuss the middle mass of the classroom (between the first and third quartile). Students often used the full range (1–100), so the extremes do not give much information. We have shown the medians in Table 1.

The rigorous CS was on average clearly less popular then the usual ICT from the previous year. The declared reasons for that are discussed above. While it has its influence, our data shows that difficulty is not a dominant factor in determining popularity or interest into a subject. However, other factors can also be linked to interest when looking at the other subjects and students' comments from the qualitative part: Generally speaking, having fun learning implies interest (according to students; any actual causality is probably trickier). Lack of usefulness on the other hand implies lack of interest.

Table 1. Median reported scores for attributes and subject groups

	Popularity	Interest	Usefulness	Mastery	Diff. of mastery	Diff. of grade
Mathematics	60	50	90	70	50	40
Sciences	50	50	60	70	50	40
CS (pilot)	**50**	**60**	**60**	**70**	**70**	**30**
ICT (last year)	80	52	50	100	10	1
Languages	70	65	99	75	20	50
Humanities	55,5	75	60	80	18	40

The difficulty to master the subject has risen enormously (from 10 to 70). In fact, our subject was clearly regarded as the most difficult to understand of all. This is interesting especially in combination with other aspects: while being the most difficult, it was still considered more interesting and more useful than standard ICT.

Another important note regards mastery. Despite being the most difficult subject to understand, our students declared they have mastered it relatively well (70, that is the same or better than mathematics, biology or chemistry). There were some who did not feel confident in CS, but the majority evaluates their mastery higher than 50. Our grades confirm this. It shows that students are more capable than it is generally thought and that they probably do not use their potential fully in other subjects.

Looking at the data for all the subjects, we may conclude: In the tracked attributes, CS fits quite well among mathematics and sciences.

6.4   What Is "informatics"

We asked students to "explain what is informatics in one sentence" during the introductory lesson. The absolutely dominant conception was about using applications, searching the internet and creating documents. That is in accord with their school experience and also not unusual, see [10, 13, 17].

In the final questionnaire, the most frequent answer to the same question was some variation on "the science about computers". 22 students consider informatics to be connected to technology strongly enough to mention it in their explanation. Another strong motive is investigating processes (most often "in computers and software", but not only). This is in accord with remarks from elsewhere in the questionnaire, that informatics investigates how and why do "things really work inside" (implying that this is specific for CS). Some students emphasize the use of mathematics, logic or thinking in general. Others emphasize the general problem-solving nature of informatics. These aspects are mentioned on various places of the questionnaire.

Data show an obvious shift in the conception of "informatics" in students. From using ICT at the beginning of our intervention, vast majority of them recognizes the deeper and more general meaning. 22 answers call informatics a science (in contrast with none at the beginning of the course). It is even more interesting considering that outside our lessons and homework, students were exposed virtually exclusively to the ICT conception.

6.5   Why Is CS Necessary in General Education

We conclude this section with broader implications. We will discuss how useful CS is in the eyes of students, and how special it is compared to other subjects. This is motivated by two common lines of argumentation against CS in general education: "It is only useful for specialists" and "Thinking and problem solving is covered in existing subjects". However, students find that CS teaches something both useful and special.

Students have clearly identified various benefits of studying CS. They consider it more useful than the usual ICT and also than biology, chemistry and a few other subjects. From sciences, only mathematics and physics are considered more useful. This is an impressive result, since all other science subjects have more time in the weekly schedule, appear during several years and are generally recognized as proper subjects. Advantages mentioned by students include efficient behavior in general, efficient problem solving, abstract, logical, "different" thinking, ability to formulate clear and quality instructions. Ability to solve problems permeates the answers in various form fields, including the question *"What makes informatics unique among other school subjects?"*

When seeking similarities, students find links in the use of numbers and logic, thus relating CS to mathematics and sciences. They also see our relation to computers, and thus similarity to the last years ICT. However, this relation is perceived as weaker. Quantitative expression of subject similarities to CS shows a clear order: the most similar subject to CS is mathematics, with median at 70. It is followed by last years ICT, with median at 50 and a surprisingly symmetric and broad distribution with quartiles at 15 and 95. The next is physics (med. at 35), then chemistry and biology.

We may conclude that students recognize special qualities in CS which they do consider important and which they do not see in other subjects. The strongest difference they see lays in focus on rigorous thinking and problem solving.

7   Methodological Recommendations

We do not see such small-scale, long-term studies very often, so we decided to include a few recommendations based on our experience.

- To get unbiased results, it must be clear to the students (and teachers) that e.g. "I do not see the use of this topic" is considered a valuable feedback, not a personal insult. Students' answers must not influence their grades.
- Using the results to apparently improve the lessons quality is an important motivational factor for sincere participation on such evaluations.

– Group interviews are an important, but potentially also misleading tool. Satisfied, introvert and other students might not be motivated enough to express their views. It is important to assess the overall opinions in the group per individual student, e.g. in a survey.
– It seems to be fruitful to combine different approaches. Interview can help to focus the questionnaire to the important topics, and the questionnaire can provide overheard motives worthy to follow up on.
– Answers such as "CS is not very useful" are rather meaningless without context. Is it more useful than last year, is it more useful than certain other subjects? Does the student consider any subjects useful? Has the student already decided about some future career? Context makes answers truly informative.

8   Conclusions

We have designed a comprehensive introductory CS course for general secondary education and piloted it in a regular classroom with 32 approximately 16 years old students within one school year. The course aimed to introduce computer science as a rigorous scientific subject and covered some most fundamental aspects of it (at an appropriately basic level). As a part of this experiment we assessed students' views toward the subject and their causes. The results of this survey are the core of this paper. We have described the course plan, the piloting conditions, the data collection and evaluation process and the results.

It turned out that students coped with the advanced topics and recognized some advantages, such as improvement in their problem solving skills or efficient behavior. They consider CS challenging, partially because they just have not been used to the necessary way of studying, working and thinking. However, students still found the subject interesting and useful, even in comparison with and other subjects and the usual "informatics" in Czechia (aiming at digital skills). Our students' conception of CS went through a significant shift towards a more appropriate idea of a science which involves logical and abstract thinking, efficient solutions to wide range of problems and creating and using technology as a tool to save resources.

Students' attitudes serve as yet another puzzle piece to show that rigorously constructed CS education could and should be a part of general education. Students can not only cope with it, they can also appreciate its benefits and regard it as a decent subject on par with mathematics and sciences.

References

1. Framework Education Programme for Secondary General Education (Grammar Schools). Výzkumný ústav pedagogický v Praze, Praha (2007)
2. Bell, T., Witten, I.H., Fellows, M., Adams, R., McKenzie, J.: Computer science unplugged: an enrichment and extension programme for primary-aged children. No. December (2006). http://csunplugged.org/
3. Bruner, J.S.: The Process of Education, vol. 115. Harvard University Press, Massachusetts (1977)

4. Dagiene, V., Stupuriene, G.: Bebras a sustainable community building model for the concept based learning of informatics and computational thinking. Inform. Educ. **15**(1), 25–44 (2016)
5. Hromkovič, J., Steffen, B.: Why teaching informatics in schools is as important as teaching mathematics and natural sciences. In: Kalaš, I., Mittermeir, R.T. (eds.) ISSEP 2011. LNCS, vol. 7013, pp. 21–30. Springer, Heidelberg (2011). doi:10.1007/978-3-642-24722-4_3
6. Hubwieser, P.: The Darmstadt model: a first step towards a research framework for computer science education in schools. In: Diethelm, I., Mittermeir, R.T. (eds.) ISSEP 2013. LNCS, vol. 7780, pp. 1–14. Springer, Heidelberg (2013). doi:10.1007/978-3-642-36617-8_1
7. Klopfer, L.: Evaluation of learning in science. In: Bloom, B., Hastings, J., Madaus, G. (eds.) Handbook of Formative and Summative Evaluation of Student Learning. McGraw-Hill, London (1971)
8. Lessner, D.: Jak žáci gymnázia vnímají výuku informatiky jako vědy. In: Didinfo 2013 - Zborník príspevkov. Univerzita Mateja Bela, Banská Bystrica (2013)
9. Lessner, D.: Informatika pro každého (2015). http://ksvi.mff.cuni.cz/ucebnice
10. Lessner, D., Vaníček, J.: Bobík učí informatiku. Matematika - Fyzika - Informatika **22**(5), 374–382 (2013)
11. Osborne, J., Simon, S., Collins, S.: Attitudes towards science: a review of the literature and its implications. Int. J. Sci. Educ. **25**(9), 1049–1079 (2003)
12. Potter, J., Wetherell, M.: Discourse and Social Psychology: Beyond Attitudes and Behaviour. Sage Publications Inc., London (1987)
13. Rambousek, V.: Výzkum informační výchovy na základních školách. Koniáš, Plzen (2007)
14. Schulte, C., Magenheim, J.: Novices' expectations and prior knowledge of software development. In: Proceedings of the 2005 international workshop on Computing education research - ICER 2005, New York, USA, pp. 143–153. ACM, New York, October 2005
15. Schwill, A.: Fundamental ideas: rethinking computer science education. Learn. Lead. Technol. **25**(1), 28–31 (1997)
16. Strauss, A.L., Corbin, J.M.: Basics of Qualitative Research, 2nd edn. Sage Publications Inc, Thousand Oaks (1998)
17. Taub, R., Armoni, M., Ben-Ari, M.: CS unplugged and middle-school students' views, attitudes, and intentions regarding CS. Trans. Comput. Educ. **12**(2), 8: 1–8: 29 (2012)
18. Wing, J.M.: Computational thinking. Commun. ACM **49**(3), 33–35 (2006)
19. Zahorec, J., Haskova, A., Munk, M.: Quality of education in programming in results of students' rating. In: Proceedings of WSEAS International Conference. Recent Advances in Computer Engineering Series, pp. 177–182. WSEAS, Paris (2012)

Typifying Informatics Teachers' PCK of Designing Digital Artefacts in Dutch Upper Secondary Education

Ebrahim Rahimi[1]([⊠]), Erik Barendsen[2], and Ineke Henze[3]

[1] Radboud University, Nijmegen, The Netherlands
e.rahimi@cs.ru.nl
[2] Radboud University and Open University, Nijmegen, The Netherlands
e.barendsen@cs.ru.nl
[3] Delft University of Technology, Delft, The Netherlands
f.a.henze-rietveld@tudelft.nl

Abstract. This paper reports on the results of the first phase of an ongoing research project in design-oriented education in informatics in Dutch upper secondary education. Our study focused on eliciting and categorizing the pedagogical content knowledge (PCK) with respect to design of digital artefacts of the informatics teachers participating in the research project. Our results suggest that teachers' PCK on design can be typified in terms of two aspects, namely (i) teachers' knowledge about objectives and goals of designing digital artefacts by students, and (ii) teachers' knowledge about ways to assess students' understanding and performance. As to (i), we distinguish an orientation towards more conceptual objectives, and one towards more practical objectives. Also with respect to (ii), we found two types of teachers' knowledge, one focused on process-based assessment and another on product-based assessment.

Keywords: Pedagogical content knowledge · Informatics education · Design education · Secondary education

1 Introduction

Design-oriented education is a well-established instructional approach to teaching informatics in secondary education in the Netherlands. As a general approach, informatics teachers define and follow different sorts of individual or group-based projects for designing and developing digital artefacts in different forms including software applications, algorithms, web sites, games, videos, podcasts, etc. [4,18]. These design projects are meant to act as a vehicle for learning, by providing opportunities for application, making errors, iteration, testing, revising and refining of newly developing conceptions and solutions, reflection, communication, representation, decision making and collaboration [13].

Design-oriented education in informatics has several theoretical and practical underpinnings. From the theoretical perspective, the learning implications

© Springer International Publishing AG 2016
A. Brodnik and F. Tort (Eds.): ISSEP 2016, LNCS 9973, pp. 65–77, 2016.
DOI: 10.1007/978-3-319-46747-4_6

and advantages of designing has been widely recognized. Design activities and challenges might lead to the creation and establishment of a participatory and collaborative learning environment around the under construction projects [2]. Through the lens of the constructionist learning theory [17], these "participatory learning environments support learners" building of understanding through the collaborative construction of an artefact and sharable product' [2] (p. 77). It is also known that design education can contribute to the understanding of scientific concepts [13], but this side of design education is hardly exploited. From a practical perspective, the learning-by-making strategy is consistent with the epistemological view of informatics as an engineering discipline [6].

This study is a part of a three-year research project called *Formative assessment of conceptual development in design education* in the context of Dutch secondary education. This project has been inspired by the new chemistry and informatics curricula in secondary education in the Netherlands [1,5] both stressing conceptual learning and design. In this project teachers and researchers from informatics and chemistry participate and collaborate in a joint research. The rationale behind combining informatics and chemistry is to support cross-fertilization of the design-based and conceptual learning approaches which seem to form the core educational activities in these subjects, respectively. The participants in the project include four researchers from Radboud University (responsible for the informatics part) and Delft University of Technology (responsible for the chemistry part), and a consortium of 12 schools/teachers (6 teachers for each part). The main purpose of this joint project is to develop appropriate assessment instruments to monitor the conceptual development of students during design activities, and to investigate the teacher knowledge required to implement design education for conceptual learning. To this end, during this project design-oriented teaching and test materials for authentic design scenarios in both chemistry and informatics will be developed and tested. The development process of these materials is combined with an investigation of the development of the teachers' PCK on the concepts to be learned and the PCK on designing digital artefacts (hereafter referred to as PCK on design).

As the first step in this joint project, we captured, described and typified the informatics teachers' PCK on design. We will use the results of this study to construct an analytical framework meant to scrutinize the design practices of the participants in the next phases of the project. Also, the results will direct the required professional development plans for the participants.

2   Pedagogical Content Knowledge (PCK)

The PCK concept has been introduced by Shulman [21] and refers to "the ways of representing and formulating the subject that make it comprehensible to others" (p. 9). A critical feature of teachers' PCK is their strategic knowledge or 'pedagogical know-how'. This strategic knowledge describes the processes that teachers follow and employ in response to the challenges of teaching specific subjects to particular learners in specific settings [22]. From a socio-cultural perspective, PCK embodies a type of teachers' professional knowledge focusing on effective and flexible transformation of subject-matter knowledge in the communication process between teachers and learners during classroom practices. PCK is integrally and inherently situated in the everyday practices of teachers and not only residing in individuals but also is distributed in their surrounding environment including books, tools, and their communities [8,10].

Capturing PCK from teachers is a difficult and challenging task. One reason for this stems from the complex nature of PCK and ways it develops. Indeed, PCK represents a personal and often tacit knowledge seldom explicitly shared between teachers [3,15] and developed and shaped after years of experience in teaching a topic [21]. The development of PCK proceeds through a non-linear, iterative and constructive process where new information is integrated with prior experiences, knowledge and beliefs captured from various domains, practices, and interactions [10].

Several models and instruments have been proposed for investigating teachers' PCK of a specific topic including: the PCK model of Magnusson et al. [16], the Content Representation (CoRe) instrument [14], the reformulation of PCK by Grossman [9], and the teacher professional knowledge and skill model [7]. Following [16], we consider four elements of teacher's PCK on a given topic: knowledge about learning goals and objectives connected to the topic (M1), knowledge about students' understanding (M2), knowledge about instructional strategies (M3), and knowledge about ways to assess students' understanding of the specific topic (M4). The Content Representation (CoRe) instrument [14] captures the key ideas connected to a specific topic, and elicits the teachers' knowledge about each idea through 8 questions. These questions cover the above four aspects of PCK. Grossman's reformulation of PCK relates it to these key questions: why to teach a specific topic? what to teach? learning difficulties associated to the topic? and how to teach the topic? The teacher professional knowledge and skill model [7] introduces amplifiers and filters to the PCK model as influential factors in amplifying or filtering teacher's learning and practice. Teacher beliefs, orientation, prior knowledge or experience, and contextual variables might serve as amplifier or filter for teacher's learning [7].

Traditionally, the PCK concept has been introduced and investigated mainly by scholars and practitioners in the context of science education. Using PCK for eliciting and portraying teachers' knowledge in computer science is an upcoming approach (e.g., [3,11,19,20]). The results of these few studies emphasize the fruitfulness of the PCK approach to investigate professional knowledge of informatics teachers [3]. In this study we aim to use the PCK concept to investigate the nature of knowledge the informatics teachers hold and utilize to support their students in their design projects. Although, the focus of the PCK concept is on capturing teachers' pedagogical knowledge in a specific topic, however, we argue that the nature of design-supporting knowledge held by teachers is a practical knowledge in a specific domain and can be captured using the PCK approach.

3   The Study Setting

The participants of this study were a group of six enthusiastic and experienced informatics teachers participating in the informatics part of the "formative assessment in design education" project. Table 1 gives the relevant information about the participants.

Table 1. The informatics teachers participating in the research

Teacher	Gender (age)	Education	Teaching (subject: duration) and other relevant experience
1	F (44)	Informatics (MSc), physics (BA)	Informatics: 8 years, Software engineer
2	M (62)	Informatics (BA), primary education (BA)	Informatics: 12 years, Dutch language: 20 years
3	M (48)	Informatics (MSc)	CS and mathematics: 20 years, university lecturer
4	M (59)	Language	Language (Dutch, English): 18 years, Informatics: 18 years, Developing help files for companies, chess player
5	M (60)	Sport	Sport, math, economics: 18 years, Informatics: 16 years, Network building experience
6	F (56)	Language	Language, Informatics

Two below research questions directed this study:

Q1: How can informatics teachers' PCK on design be described?
Q2: Which parameters could be used to categorize informatics teachers' PCK on design?

Due to the exploratory nature of this study, we chose qualitative research methods for data collection and analysis. Given its in-depth and exploratory approach, we selected the interview as the main method to collect data (cf. [10,15]). We used the four constituting elements of PCK (i.e., M1, M2, M3, M4) [16] together with a combination of CoRe questions [14] and Grossman's [9] questions to construct a set of interview questions to elicit teachers' PCK on design, see Table 2.

Six individual semi-structured interviews were conducted with the participants. Five of the interviews were conducted in the participants' schools and one interview took place at Radboud University. Each interview lasted about two hours. All interviews were recorded using a voice recorder for further analysis. The collected data then were analyzed by the research team. The analysis procedure included transcribing audio data verbatim, coding data, reading the

Table 2. The interview questions for eliciting informatics teachers' PCK on design

PCK elements ([16])	Questions about CS design projects (adapted from [14] and [9])
M1. Knowledge of goals and objectives	1. Why do you ask your students to do software projects in your CS courses?
	2. What do you like/not like about software development projects by your students?
M2. Knowledge of students' understanding and practices	3. What sorts of skills do students need to acquire in order to be able to develop software?
	4. What are the learning difficulties/problems concerned with the software development projects in your classrooms?
	5. What do students actually learn from their software development projects?
M3. Knowledge about instructional strategies	6. What to teach students to achieve the project development objectives?
	7. How to teach students to achieve the project development objectives?
	8. What are the teaching difficulties/problems concerned with the software development projects in your classrooms?
	9. What technological tools do you use in your classrooms?
M4. Knowledge about ways to assess students' understanding	10. How do you assess your students' learning and achievement during their project development experiences?

transcripts organized by codes, writing memos, recoding and merging similar codes as necessary, grouping codes into categories, reviewing and confirming codes by all the research members, and writing up conclusions.

We used the teacher professional knowledge and skill model [7] in addition to the four PCK elements [16] as the analytical framework for coding the data and investigation the relationship between the emerged codes. Furthermore, to code the technology knowledge of the participants we borrowed some codes from the Technological Pedagogical and Content Knowledge (TPACK) model [12].

4   Results

In this section we use the processed data to answer the first research question and describe the informatics teachers' PCK on design and its amplifiers and filters.

Knowledge About Goals and Objectives (M1). We concluded that the informatics teachers' knowledge about objectives of design contains 10 objectives that can be divided into three main categories:

(i) *Conceptual objectives:* this category of objectives emphasize the importance of developing digital artefacts as a means for learning and understanding computer science (CS) concepts. The identified conceptual objectives for design projects include: *learning CS concepts, realizing students' knowledge gap, acquiring programming knowledge and skills,* and *incorporating theory and practice* (emphasized mainly by the teachers 1, 2, 3, 4).

(ii) *Motivational objectives:* this category emphasizes the 'fun' and motivational aspects of design practices to trigger students' learning and engagement. This category consists of three objectives, namely: *recognizing and addressing students' differentiation, motivation and preparing students for ICT-based jobs and subjects,* and *making a workable product (mainly for real customers) as a means for making students learning tangible and also touch their feeling of accomplishment, ownership, and sharing* (asserted by all participants).

(iii) *Practical objectives:* the focus of this category is mainly on the practical benefits and advantages of developing digital artefacts. It involves three objectives: *acquiring soft and design skills* (i.e. problem solving, communication, collaboration, design thinking, etc.); *experiencing real world problems, challenges, and way of thinking; becoming an independent learner;* and *learning about the latest ICT developments and trends* (emphasized mainly by the teachers 4, 5, 6).

Knowledge About Students' Understanding and Performance (M2). We divided teachers' knowledge about students' understanding and practice that influences teacher's instruction into eight categories as follows:

(i) *Students' faced problems:* according to the interviewees, students experience different sorts of problems during their design practices, including: group issues (i.e., free riding, peer assessment), technical problems, orientation/planning problems, difficulty in finding real case projects, superficial/ shallow learning mainly due to following a non-reflective approach to designing by students, difficulty in understanding the semantic of a problem, inability to transfer their theoretical knowledge into action to solve real problems (i.e. inability to breakdown a problem and using CS concepts such as loop to solve it or difficulty in generating appropriate algorithms). Addressing these problems significantly shape and influence the participants' instruction.

(ii) *Development of students' soft and design skills:*referring to teacher's knowledge and understanding of the level and development of soft and design skills in students.

   "In a game making project, my students experienced several problems about social skills and customer relationship. Accordingly, I decided to

change my plan for projects to define two extra roles in our project, being: manager (undertaken by the teacher) and customer to help students to learn appropriate social skills by observing the manager (teacher) communication by the customers" (Teacher 3).

(iii) *Students' learning process:* representing teachers' knowledge of the specifications of students' learning process including their learning and design goals, planning, activities, intermediate products, faced problems, taken solutions, revisions, etc. (remarked by the teachers 2, 3, 4). To capture this type of knowledge (M4), the participants use different approaches such as tracing and analyzing students' log book Teacher 4), observing students' practices, or using SCRUM methodology (Teacher 2).

(iv) *Students' understanding of their projects' structure:* referring to the teachers' knowledge about students' understanding of their projects' purposes, structure, content, and concepts underpinning their products.

(v) *Development of students' conceptual understanding:* meaning teachers' knowledge and awareness on students' conceptual knowledge and knowledge gaps. This type of M2 knowledge has an inviable position in shaping and influencing participants' instructional activities (M3), as expressed by one of the teachers:

"After analyzing their log files, I realized many students cannot calculate the average of a list of numbers in SQL. Thus, I decided to adjust my teaching materials and teach lessons about AVG and other mathematical functions in SQL" (Teacher 2).

(vi) *Students' reaction and perception:* entailing teachers' knowledge and awareness of students' reaction on and perception of design projects. The influence of this understanding on teachers' instruction (M3) is shown below:

"As an educational system you have to compete with other systems that make learning fun for children. The social components of classrooms are the most competitive advantage and favourite part of the school activities for students. We should invest on these social components to make schools fun and meaningful for students" (Teacher 5).

(vii) *Students' preferences and orientation:* referring to teachers' knowledge of students learning preferences and orientations. Capturing this type of knowledge is essential for recognizing and addressing differentiation between students (M3) as a educational principle promoted by many schools:

"When I started introducing and using Appinventor in my course, there were a lot of students who liked to be hacker and know about cyber security. Accordingly, we made some different modules where they could choose what they liked to learn" (Teacher 5).

(viii) *Students' level of performance in their projects:* referring to teachers' knowledge about the activeness of students in their group projects. Teachers use this type of knowledge to grade, trigger and encourage students to actively participate in their group projects. The participants mentioned different approaches to achieve this knowledge ranging from direct observation of students working (the teachers 1, 2, 4, 5, 6) to asking students themselves to rank their peers' level of performance and activeness (Teacher 3).

Knowledge About Instructional Strategies (M3). We identified and divided the teachers' knowledge about instructional strategies associated with conducting design projects into 8 categories as described below:

(i) *Project development and management:* referring to teachers' knowledge about shaping the students' activities according to the respective phases of software development, and teachers' skill and ability to manage and scaffold students to construct their projects and achieve their project's objectives.

(ii) *Linking conceptual content:* meaning teachers' knowledge and skills to evaluate, develop and update content required by individual students for developing their projects.

(iii) *Digital resources:* implying teachers' knowledge and ability to use technology to provide new ways of teaching CS concepts. For example, Teacher 1 uses the `code.org` service to teach complex CS and programming concepts through providing simple examples.

(iv) *Digital tools:* referring to teachers' knowledge about the affordances and constraints of technology as an enabler of different teaching approaches, that is, technology enhanced learning. For example, Teacher 4 uses the *itslearning* learning management system to log, monitor and trace the learning process of students.

(v) *Stimulating student-centric, flexible, differentiated and collaborative learning:* referring to teachers' knowledge and ability to implement and support student-centric, differentiated and collaborative learning scenarios. This sort of knowledge is embodied in different instructional activities of the participants including involving students in choosing their project subjects, implementing peer review activities, asking and encouraging students to participate in generating teaching materials and even defining final exams' questions (Teacher 5), supporting flexible learning by providing students with learning choices and allowing them what, when, where and how to learn (the teachers 1, 4), promoting differentiated learning through teaching various content and providing separate assignments for different students (Teacher 1), encouraging students to reflect on the structure of their projects and their individual and group performance (the teachers 2, 3, 6).

(vi) *SCRUM-based project development:* referring to teachers' knowledge and ability to implement SCRUM methodology in their classrooms (Teacher 2).

(vii) *Customer-students relationship management:* referring to teachers' knowledge about the management of relationship between customers and developers of projects.

(viii) *Drama:* referring to teachers' knowledge and ability to implement drama in their classrooms to ease teaching CS concepts (Teacher 3).

Knowledge About Ways to Assess Students' Learning and Performance (M4). We described teachers' knowledge about ways to assess students' understanding and performance using these codes: *final and intermediate design products, presentations, tests* (traditional or automated tests), *short quizzes, daily check questions, assignments, customers' feedback on developed products, peer assessment, students' reports, students' log book, teacher's observation, discussion.* Among these assessment tools, customer's feedback represents a new approach being used by the teachers 2, 3, 4, 5. Diverse approaches have been followed by these teachers to choose projects' customers, including: defining real customers (by the teachers 2 and 4), defining another teacher to play the role of customer (Teacher 3), and playing the role of customer by teacher himself (Teacher 5). Interestingly, while four teachers see significant learning benefits in customer's feedback, Teacher 6 does not follow this approach as she believes finding appropriate customers with realistic expectations consistent with students' knowledge and level of expertise is difficult.

5   Elaboration

In this section we elaborate on the aforementioned results to find parameters that can be used to categorize informatics teachers' PCK on design. Two highlights of informatics teachers' PCK on design can be inferred from the results. First, with regard to teachers' knowledge about objectives and goals of design, a diverse set of objectives have been identified that direct the design-based instruction of the participants. These objectives form a continuum, shown in Fig. 1, consisting of 10 objectives ranging from more conceptual objectives on one side to more practical objectives on the other side.

One reason for this diversity in objectives stems from the flexible, less structured and teacher-dependent characteristics of informatics education in the

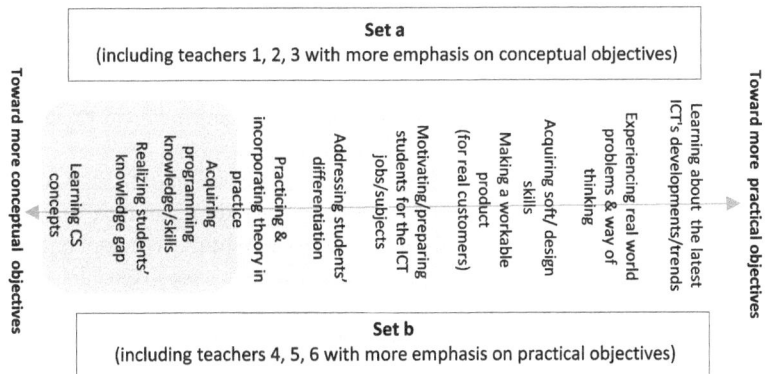

Fig. 1. The continuum of teachers' knowledge of objectives and goals of design projects

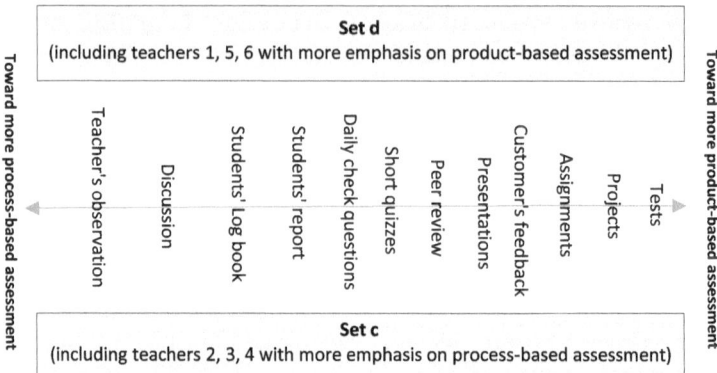

Fig. 2. Ways used by participants to assess students' understanding and performance

Netherlands. This categorization has led to the emergence of two sets of teachers, namely a set **a** consisting of the teachers 1, 2, 3 with more emphasis on conceptual objectives, and a set **b** of the teachers 4, 5, 6 with more emphasis on practical objectives. Interestingly, as described in Table 1, the teachers in set **a** have an informatics related educational background, while the teachers in set **b** have a non-informatics related educational background. This relation resembles the observations by Barendsen et al. concerning the perceived learning objectives of programming between teachers with informatics-related education and other teachers [3].

The second highlight of the teachers' PCK on design concerns the teachers' knowledge of ways to assess students' understanding and performance. As described earlier, the participants have knowledge about diverse ways to assess their students' learning and performance. These ways form a continuum ranging from more process-based assessment (i.e. observation and discussion) on one end to more product-based assessment (i.e. final products, assignments, tests) on the other end, as shown in Fig. 2.

Two sets of teachers can be discerned on the basis of this continuum: set **c** involving the teachers 2, 3, 4 with more emphasis on process-based assessment, and set **d** consisting of the teachers 1, 5, 6 with more emphasis on product-based assessment.

Based on these two PCK elements, we can typify teachers' PCK on design. Combining the elements M1 and M4 results in the identification of four types of informatics teachers' PCK, as shown in Fig. 3, namely: conceptual-product-based PCK (mainly held by Teacher 1), conceptual-process-based PCK (mainly held by the teachers 2 and 3), practical-product-based PCK (mainly held by the teachers 5 and 6), and practical-process-based PCK (mainly held by Teacher 4). These four types of teachers' PCK can be understood as representing teachers' individual orientation toward design-based education in informatics that support or shape their design-based instruction:

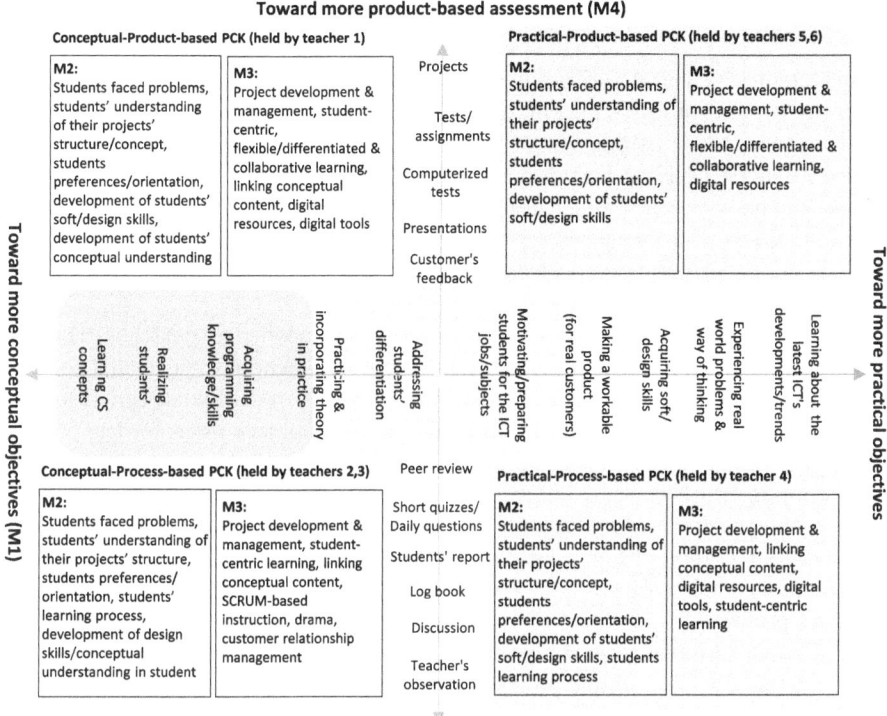

Fig. 3. A model to categorize informatics teachers' PCK on design

- *Conceptual-product-based PCK* represents teacher's orientation toward: recognizing and fulfilling more conceptual objectives for design projects (element M1) and understanding students' development of conceptual learning (M2) through product-based assessment approaches (M4) and linking conceptual content, digital resources, and digital tools (M3).
- *Conceptual-process-based PCK* refers to teacher's emphasis on: addressing more conceptual objectives (M1) and understanding students' development of conceptual learning (M2) through mainly process-based assessment approaches (M4) and SCRUM-based or drama instructional strategies (M3).
- *Practical-product-based PCK* represents teacher's orientation toward: fulfilling more practical objectives (M1) and understanding students' development of soft and design skills (M2) through mainly product-based assessment approaches (M4) and rich knowledge of digital resources (M3).
- *Practical-process-based PCK* refers to teacher's emphasis on: addressing more practical objectives (M1) and understanding students' development of soft and design skills (M2) through mainly process-based assessment approaches (M4) and rich knowledge about digital resources and digital tools (M3).

6   Conclusion and Discussion

In this paper we elicited and typified the PCK on design of six informatics teachers in the context of upper secondary school in the Netherlands. The results suggest that two distinguishing aspects of teachers' PCK can be used to typify informatics teachers' PCK on design, namely, their knowledge of objectives and ways of assessment. By combining these two elements, a model has been identified with four types of informatics teachers' PCK on design. This model represents teachers' orientations toward design that support or direct their design-based instruction in the classroom.

The model could serve to provide insight on knowledge patterns, themes, differences, and similarities among the informatics teachers with regard to their design-based instruction. The provided insight has a multi-folded functionality. First, it might inform the required professional development plans for the participating teachers. Moreover, the typified knowledge might be used as an analytical and planning framework to analyse, scrutinize, and prescribe the design practices of the teachers. Finally, we expect the results to provide us with good practices and evidence of empirical and contextualized design principles, leading to a model needed to direct the development of the course and test materials in our joint project.

The small number of participants can be seen as a limitation of the study. However, the diversity among our teachers with respect to the PCK elements M1 and M4 appears to resemble the variation in the practice of Dutch informatics teachers found in a wider study with 178 informatics teachers covering 59 percent of the population of informatics teachers in the Netherlands [4]. Whether the classification into four PCK types will still hold in the larger population, is to be investigated.

It is known that amplifiers and filters influence teachers' PCK development, see Sect. 2. Our data appeared to be sufficiently rich for an in-depth analysis of these influencing factors. We will report on this in a later paper.

References

1. Apotheker, J.H.: Developing a new chemistry curriculum in The Netherlands. Abs. Pap. Am. Chem. Soc. **237**, 1155 (2009)
2. Barab, S.A., Barnett, M., Yamagata-Lynch, L., Squire, K., Keating, T.: Using activity theory to understand the systemic tensions characterizing a technology-rich introductory astronomy course. Mind Cult. Act. **9**(2), 76–107 (2002)
3. Barendsen, E., Dagienė, V., Saeli, M., Schulte, C.: Eliciting computer science teachers' PCK using the content representation format: experiences and future directions. In: Gülbahar, Y., Karataş, E., Adnan, M. (eds.) Proceedings of the 7th International Conference on Informatics in Schools: Situation, Evolution and Perspectives (ISSEP 2014), pp. 71–82 (2014). Selected Papers
4. Barendsen, E., Fisser, P., Krüger, J., Tolboom, J.: Herziening van het Nederlandse informaticacurriculum havo-vwo Paper presented at ORD 2014, Groningen (2014)

5. Barendsen, E., Mannila, L., Demo, B., Grgurina, N., Izu, C., Mirolo, C., Sentance, S., Settle, A., Stupurienė, G.: Concepts in K-9 computer science education. In: Proceedings of the 2015 ITiCSE on Working Group Reports, pp. 85–116. ACM (2015)
6. Berglund, A., Lister, R.: Introductory programming and the didactic triangle. In: Proceedings of the Twelfth Australasian Conference on Computing Education, vol. 103, pp. 35–44. Australian Computer Society, Inc. (2010)
7. Gess-Newsome, J.: A model of teacher professional knowledge and skill including PCK. In: Berry, A., Friedrichsen, P., Loughran, J. (eds.) Re-examining Pedagogical Content Knowledge in Science Education, pp. 28–42. Routledge (2015)
8. Greeno, J.G., Collins, A.M., Resnick, L.B.: Cognition and learning. In: Berliner, D.C., Calfree, R.C. (eds.) Handbook of educational psychology, pp. 15–46. Macmillan, New York (1996)
9. Grossman, P.L.: The Making of a Teacher: Teacher Knowledge and Teacher Education. Teachers College Press, New York (1990)
10. Henze, I., Van Driel, J.H.: Toward a more comprehensive way to capture PCK in its complexity. In: Berry, A., Friedrichsen, P., Loughran, J. (eds.) Re-examining Pedagogical Content Knowledge in Science Education, pp. 120–134. Routledge (2015)
11. Hubwieser, P., Magenheim, J., Mühling, A., Ruf, A.: Towards a conceptualization of pedagogical content knowledge for computer science. In: Proceedings of the Ninth Annual International ACM Conference on International Computing Education Research, pp. 1–8. ACM (2013)
12. Koehler, M.J., Mishra, P.: What is technological pedagogical content knowledge? Contemp. Issues Technol. Teach. Educ. **9**(1), 60–70 (2009)
13. Kolodner, J.L., Camp, P.J., Crismond, D., Fasse, B., Gray, J., Holbrook, J., Puntambekar, S., Ryan, M.: Problem-based learning meets case-based reasoning in the middle-school science classroom: putting learning by design (tm) into practice. J. Learn. Sci. **12**(4), 495–547 (2003)
14. Loughran, J., Mulhall, P., Berry, A.: In search of pedagogical content knowledge in science: developing ways of articulating and documenting professional practice. J. Res. Sci. Teach. **41**(4), 370–391 (2004)
15. Loughran, J., Mulhall, P., Berry, A.: Exploring pedagogical content knowledge in science teacher education. Int. J. Sci. Educ. **30**(10), 1301–1320 (2008)
16. Magnusson, S., Krajcik, J., Borko, H.: Nature, sources, and development of pedagogical content knowledge for science teaching. In: Gess-Newsome, J., Lederman, N.G. (eds.) Examining pedagogical content knowledge, pp. 95–132. Dordrecht, Kluwer (1999)
17. Papert, S., Harel, I.: Situating constructionism. Constructionism **36**, 1–11 (1991)
18. Rahimi, E.: A design framework for personal learning environments. Ph.D. thesis, Delft University of Technology, The Netherlands (2015)
19. Saeli, M.: Teaching programming for secondary school: a pedagogical content knowledge based approach. Ph.D. thesis, Eindhoven University of Technology, The Netherlands (2012)
20. Saeli, M., Perrenet, J., Jochems, W.M.G., Zwaneveld, B.: Teaching programming in secondary school: a pedagogical content knowledge perspective. Inform. Educ. **10**(1), 73–88 (2011)
21. Shulman, L.S.: Those who understand: knowledge growth in teaching. Educ. Researcher **15**(2), 4–14 (1986)
22. Shulman, L.S.: PCK: Its genesis and exodus. In Berry, A., Friedrichsen, P., Loughran, J., eds.: Re-examining Pedagogical Content Knowledge in Science Education, pp. 3–13. Routledge (2015)

Students' Success in the Bebras Challenge in Lithuania: Focus on a Long-Term Participation

Gabrielė Stupurienė, Lina Vinikienė, and Valentina Dagienė[(⌗)]

Vilnius University Institute of Mathematics and Informatics,
Akademijos Street 4, 08663 Vilnius, Lithuania
{gabriele.stupuriene,lina.vinikiene,
valentina.dagiene}@mii.vu.lt

Abstract. The paper deals with students' participation in the Bebras challenge on Informatics and Computational Thinking in Lithuania in 2010–2015. As noticed, secondary school students have an opportunity to learn the basic informatics concepts during the participation in the Bebras challenge. Analyses of a large amount of data from participants' task solving records are provided. Additionally, observation of the task difficulty level of the Bebras contest in the past 6 years is presented. The target group, on which a research study was focused, is a group of students who solved tasks 6 years in turn. A detailed overview of their results provides an understanding how the participants have solved tasks over these years. The importance of algorithmic thinking as an opportunity for students to learn and understand the basics of informatics as well as develop their computational thinking skills is emphasised. The results of data analysis highlight the importance of students' achievements by a long-term participation.

Keywords: Bebras challenge · Informatics education · Learning algorithms · Problem solving · Task difficulty · Computational thinking

1 Introduction

The Bebras challenge on Informatics and Computational Thinking continues growing and expanding into new countries [3]. Students' interest in solving puzzle-based informatics tasks, gamification, and attractiveness are the main reasons for growing. Students get introduced to informatics concepts basically without any requirement of pre-knowledge in informatics. The Bebras challenge is designed for all school students from primary education until school leaving age [10]. In Lithuania, the Bebras challenge is considered as a part of learning/teaching process. Teachers are ready to present examples of Bebras tasks when starting introductory parts to various topics of Informatics and willing to attract students' attention to the basics of informatics by solving short tasks.

Students are asked to participate in the challenge during information technology (IT) lessons: in Lithuania the IT course (including Informatics topics) is mandatory from the 5th to 10th grades. Over 530 schools participated in Lithuania in 2015.

© Springer International Publishing AG 2016
A. Brodnik and F. Tort (Eds.): ISSEP 2016, LNCS 9973, pp. 78–89, 2016.
DOI: 10.1007/978-3-319-46747-4_7

Participants were asked to answer multiple-choice questions, solve open-ended tasks, and interactive tasks. The tasks were designed with a view to influence students' understanding of informatics concepts and developing some computational thinking skills, for example, execution of algorithms, representation of data and data analysis, problem solving, modelling, abstraction of structures and processes [6].

The paper is aimed to give an overview of students' performance in the Bebras challenge in a long-term participation (up to 6 years) in Lithuania. The participation in the contest during the so-called Bebras week is a short-term activity due to the defined domain and time taken for task solving. Students have to solve tasks within 45 min; their willingness to solve tasks next year is not known. A long-term participation is indicated in the sense of six-year continuity of participation and perseverance to be a success despite the task content and difficulty.

We are going to discuss students' participation in the Bebras contest and solving tasks during 6 years (2010–2015). The average of students' scores gained during contests, difficulty levels of the given tasks, and time taken for solving each task are analysed in detail. Students' passion to solve Bebras tasks each year in turn can be considered as a successful way of learning or teaching informatics.

The research questions we address in this paper are:

1. Are students interested to participate in the Bebras contest year-by-year?
2. Did students' results improve during a long-term participation?
3. Have task difficulties influenced students' participation?

These research questions may provide understanding how the challenge is relevant to students, how they succeed, and which informatics concepts can be trained/learned while solving tasks.

2 Related Works

The Bebras challenge helps to motivate students in solving tasks, attract their attention to informatics concepts. According to White [18], learners have intrinsic motivation when they learn something new or succeed in solving tasks. Similarly, Palmer (2005) claims that motivation to learn is based on students' capabilities and activities that incite their imagination and abilities to compare the real life situation to practical task solving. Palmer (2005) claims that students are able to range the personal and performance goal. Also, he has approved, that students demonstrate their abilities when they achieve success in solving a task, but their results could be influenced by motivation to focus on other participants' achievements rather than on the content of the task. A driving motif in learning activities and achievements is interest. Interest can be classified as long-term and short-term. A long–term interest is described as the preference to a particular domain (personal interest), while a short–term interest focuses on a specific temporary situation. Furthermore, students' interest in activities is relevant to novelty, meaningfulness, and involvement of a particular domain. A short–term success is a source of self–efficacy. This success is a success "in understanding science concepts rather than in passing assessment tasks" [17]. The long–term success is supported by opportunities to practice learnt concepts. Moreover, in the teaching process a difficulty of tasks should be chosen carefully: tasks solved successfully in a few minutes; concepts selected according to the task difficulty; a task should not require a sequence of experiences before the students understand it; the learning should be performed step-by-step; the importance of creativity, fantasy, imagination, and feedback is desirable [17]. These ideas are emphasized in the teaching process and implemented in the Bebras tasks [9].

The Bebras challenge can be considered as an approach to learn informatics fundamentals. Students are able to develop some computational thinking skills and to solve tasks in a logical and systematic way.

Computational thinking is considered as a fundamental skill that includes a problem solving process, understanding of human behaviour and development of new solution methods [13, 15]. In computer science education, CSTA distinguishes concepts and capabilities such as data analysis, data collection, data representation, problem decomposition, abstraction, algorithms and procedures, parallelization, and simulation [6].

Generally speaking, students in lower grades show better results. They gain new knowledge and develop skills faster [2]. That is why new knowledge and skills should be taught gradually and practiced over a long period.

Bebras tasks developers have created tasks based on the new knowledge of informatics and intention to develop students' computational thinking. Organizers of Bebras challenge ask students to participate in the contest from early age and improve their knowledge constantly. Interactivity, visualization, correctness, solvability, independent of the curriculum are Bebras tasks attributes [4, 7, 15]. Informatics tasks should be based on the logical thinking and unconventional solution in order to develop students' computational thinking skills [2]. Interactivity and visualization are very important in the development of student imagination or creativity [17], students' ability to be creative is a source of intrinsic motivation.

Bellettini (2015) has noted that tasks, used in the challenge, are either easier or more difficult than expected by task developers. For example, a growing number of topics in the programming contest [11] is emphasized as one of the reasons of increasing task difficulty. According to Forišek (2010), participants should solve more difficult tasks each year due to the experience in learning (a practice). In the case of the Bebras challenge, task difficulty is related not only with the practice, but also with interrelation between criteria and competence requirement, complexity of task, interface peculiarities of a competition, task presentation on the screen, etc. [10]. It is an important observation, that depending on a task, "a student with a high level of thinking has the possibility but not the necessity to react on the highest possible level" [20].

Bebras tasks give a deeper understanding of algorithms used in everyday situations. Algorithms and data representation make up the main part in the Bebras tasks. Forišek (2010) mentioned that the trend of tasks, based on the computational thinking problem, required algorithms. Learning of algorithms becomes easier and more understandable due to visualization of algorithms [12]. By Futschek (2006), the possibility to try algorithms visually gives "a feeling why algorithms work and how algorithms may be improved". Participants are able to learn and design better through the visualization. Concepts of algorithms are taught in schools and play an important role in the curriculum [8, 14]. The understanding of algorithms perceives the sense of real life situation, and digital life experience [5]. For example, our social life is unimaginable without Facebook, Twitter. These platforms render a good opportunity for developing algorithmic thinking.

Interactive elements of the task (graphic, animation, etc.) support student understanding of a content and how they construct meaning from the presented content. The following goals of the interactive tasks benefit could be mentioned: greater validity, increased student engagement and motivation, measurement of higher order thinking skills, promoted students' reflection by solving tasks, better evaluating the cognitive and problem-solving skills [21]. For example, the interactivity of Bebras tasks makes the challenge more attractive if interactivity means manipulation with mouse. If tasks stem out from real life situations, it will be appreciated by older contestants. Younger will enjoy motivation through the fairy tale character of Bebras [23].

3   Data Analysis

Data from the Bebras contest are collected in 2010–2015. We have recorded the data from 131310 participants in total. The analysis is based on the overview of data of participants' grades and scores gained for each task as well as task difficulty and the time taken to solve each task. Also, the informatics concepts involved in tasks during the chosen period are discussed. The data analysis consists of the following steps:

1. Selecting data about participants of the Bebras contests from 2010 to 2015 and selection of students who participated in each challenge (6 years in turn).
2. Comparing participants' results and time taken to solve each task.
3. Reviewing tasks that were solved by the target group according to task difficulty and the average of students' scores.

Participants are divided into 5 age groups and solved from 18 to 24 tasks within 45 or 55 min. Each task has one of the three difficulty levels (easy, medium, hard) as prescribed by developers. The set of tasks consists of 6 (or 8 in the case of the 24 task set) tasks of each difficulty level.

Our target group is 137 students who participated and solved tasks in all the 6 years of Bebras contests (2010–2015).

3.1   Students Are Interested to Participate in the Bebras Contest Year-by-Year

A distribution of participants (by grades, percentages of boys and girls) in the Bebras contests in the period of 2010–2015 is provided in Table 1. The results show that the number of participants was almost stabilized, especially when keeping in mind that the number of students is declining of late years. Girls are interested in the participation as well as boys. There is a tendency that girls of lower grades (from 3rd to 8th) are more interested in the participation. For this age the percentage of girls is close to 50 %. The percentage of girls does not exceed 44 % each year in the 9th and 10th grades and is less than 33 % in the 11th and 12th grades. The declined number of girls in the 11th and 12th grades might be influenced by motivation to select the IT (informatics) exam as a

Table 1. Participants' distribution in the Bebras contest during 2010–2015.

Grade	2010			2011		
	G*	B*	T*	G*	B*	T*
3–4	-	-	-	-	-	-
5–6	43.2	56.8	3106	43.4	56.6	5306
7–8	43.2	56.8	3344	42.4	57.6	5038
9–10	39.0	61.0	3808	38.8	61.2	5561
11–12	29.4	70.6	2660	31.6	68.4	3323
Number of participants			**12918**			**19228**

Grade	2012			2013		
	G*	B*	T*	G*	B*	T*
3–4	47.6	52.4	2049	44.9	55.1	2175
5–6	45.3	54.7	6333	51.9	48.1	6210
7–8	42.9	57.1	6423	43.3	56.7	6547
9–10	39.9	60.2	6168	40.2	59.8	6485
11–12	32.1	67.9	3416	31.8	68.2	3671
Number of participants			**24389**			**25088**

Grade	2014			2015		
	G*	B*	T*	G*	B*	T*
3–4	43.3	56.7	2410	44.8	55.2	2374
5–6	38.2	61.8	6268	46.8	53.2	7100
7–8	49.5	50.5	7169	44.2	55.8	5810
9–10	43.7	56.3	5990	43.9	57.1	6114
11–12	30.4	69.6	3148	31.2	68.8	3304
Number of participants			**24985**			**24702**

*G – girls, B – boys, T – total number of participants

maturity exam (and also selection of the optional programming module related to the exam).

Slight changes in participation numbers are notable from 2014. The declining numbers (from 0.4 % to 1.13 %) of participants tend to vary due to a declined population in Lithuania (and the number of children). OECD (Organization for Economic Co-operation and Development) reports that the number of students decreased by 30.6 % in general education schools during 2005–2012 and the inhabitants' number dropped to 16 % in Lithuania from 2011 to 2014 [16].

In Lithuania school leaving students are required to take at least 3 but not more than 6 matriculation exams and IT (half based on the optional programming module) is one of them. In the 9th grade students have to make a decision on their interest domain and select the desired learning subjects. Learning IT "is aimed at summarizing and systematizing students' knowledge drawing attention to the right application of technologies and their legitimacy" [8] in the 9th and 10th grades. Additionally, there is a possibility to choose one of three optional modules: basics of programming, web design or electronic publishing. The IT curriculum of grades 5–8 emphasizes the ability to apply computers in the learning process, creativity of knowledge construction, critical thinking, self-confidence, ability to express their own view, and attitude to process data using software. Informatics teaching is implemented in after-school activities.

We have defined that 137 students participated in the Bebras contest 6 years in turn. They started to participate from the 5th, 6th, and 7th grades, respectively. 52 participants started to solve tasks from the 5th grade and participated each year, 50 participants entered at the 6th grade, and 35 participants started from the 7th grade.

A detailed overview of 137 students has showed that 44.2 % of girls from the 5th grade, 16 % of girls from the 6th grade, and 17 % of girls from the 7th grade were involved in a long–term participation. These numbers show that girls are interested in long–term task solving when they are involved from the earlier age (lower grades). Additionally, we see a tendency that boys are more interested in solving tasks and participating in contests.

3.2   Participants Are Able to Improve Their Results During a Long-Term Participation

We have observed students task solving results in the Bebras contests for many years. Students' achievements were reviewed in the following three steps:

1. Studying the results of each participant through a long–term participation (6 years);
2. Analysing how many participants solve tasks correctly using informatics concepts;
3. Comparing the score averages of the target group and students who solved the same set of tasks.

Six participants are able to solve correctly over 52.4 % of tasks during the Bebras contests in a long–term period (6 years). 10 participants are successful in 54 % of tasks in the period of 5 years. 9 out of 137 participants achieve a success in solving more than 90 % of tasks (students, who solved correctly over 90 % of tasks, achieved the highest scores).

The results of the most successful participants are presented in Table 2. The results are distributed by grades, task scores are distributed by years, and the highest scores collected by the participants who solved the same set of tasks. Note that, one participant has solved over 90 % of tasks correctly 5 years in turn and achieved the highest score 2 years in turn. Most of the participants with the best results are in grades 8–12. 4 out of 137 participants have achieved the highest score (marked in italics). Only 2 girls have solved over 90 % of the set of tasks.

Only one participant has solved the set of tasks better each year in a long–term period. He was solving correctly about 5 % more of tasks each year. 44 students solved tasks correctly in the period of several years. 35 of them solved tasks better each year in the period of 3 years, 7 participants solved tasks successfully 4 years in turn and 1

Table 2. Participants who have solved correctly more than 90 % of tasks.

Year	Grade	Correctly solved tasks	Gender	Scores collected by participants of the target group	Highest score in a set of tasks
2013	8	95.2 %	Female	83.75	100
2013	8	90.5 %	Male	83.75	100
2014	9	94.4 %	Male	208	216
2012	7	100 %	Male	100	100
2014	10	94.4 %	Female	200	216
2011	7	91.7 %	Male	112.5	115
2012	8	95.8 %	Male	116.25	120
2013	9	95.2 %	Male	100	100
2014	10	100 %	Male	216	216
2015	11	94.4 %	Male	200	216
2015	12	94.4 %	Male	200	216
2014	11	100 %	Male	216	216
2010	7	95.8 %	Male	116.25	116.25

participant got better results 5 years in turn. Most of these students tried to solve more tasks correctly in grades 6–9. The results of other students are different each year (higher or lower than in the previous year).

The number of correct answers distributed by difficulty of the tasks were analysed in detail. The data of participants who started to solve tasks in the 5th grade are demonstrated in Table 3. It is obvious how the number of correct answers is distributed by the task difficulty provided by developers of tasks.

The tasks introduce algorithmic problems each year. More students solve the tasks correctly each year, except 2014 and 2015 (grade 9–10). The same situation is with the percentage of correct solutions in groups of participants who solved tasks from the 6th

Table 3. The percentage of students who solved tasks, focussed on algorithms, correctly

	Algorithms		
Year	Easy	Medium	Hard
2010	37.5	30.77	29.81
2011	59.62	46.63	31.73
2012	62.5	40.87	31.54
2013	67.95	58.33	53.85
2014	15.38	56.73	19.23
2015	57.69	35.26	32.05

and 7th grades 6 years in turn, respectively. The tasks which introduced algorithms were solved by all the participants each year. The percentage of correct solutions is decreasing from the 9th grade. The participants are not able to solve "hard" tasks very well in grades 9–12 (solvability of tasks decreased up to 19 %).

Participants tried to achieve a success searching for the best solution and to improve their results.

The score averages of participants were calculated in each year contest (Table 4). Due to the score variety, the scores of participants were normalized up to 100 according to the highest scores, collected by students in the respective grade.

Table 4. The score averages of participants distributed by years.

Year	A[a]	AA[b]	B[a]	BB[b]	C[a]	CC[b]
2010	41.58	40.45	53.35	43.71	47.61	41.89
2011	43.11	38.60	52.8	37.61	49.26	41.26
2012	41.39	35.35	50.85	38.11	60.66	37.82
2013	53.26	45.41	47.05	35.87	51.05	39.86
2014	44.87	37.12	54.52	39.96	53.98	48.19
2015	45.16	39.15	59.20	43.87	55.23	48.97

[a]A, B, C – averages of students who started to solve tasks from grade 5, 6, 7, respectively.
[b]AA, BB, CC – averages of all students who solved tasks at the same time as the students from the 5th, 6th, 7th grades, respectively.

It is evident, that the participants who have solved tasks in a long–term period are able to achieve better results. They collect higher scores than the average of groups. They tried to improve their knowledge because the score average is slightly growing through the year. The participants tend to have higher results in the 11th and 12th grades (over 50 scores), – we can say that they are more motivated to learn informatics and probably have chosen the optional programming module or participated in extracurricular informatics activities.

3.3 Value of Task Difficulty

A further analysis of the value of task difficulty is needed in order to find a relation between the participants' success in answers and task difficulty, provided by developers. This analysis is necessary because not all the participants are able to solve more than 50 % of tasks in the set correctly. Besides, there are some tasks that can be solved correctly only by a small part of participants.

The value of difficulty of each task was calculated. A calculation of the task difficulty value involves all participants' abilities to solve the task. The value of difficulty is considered as a ratio between the number of correct answers and the total number of answers (the number of tasks that students have not tried to solve at all). Lower values indicate more difficult tasks and higher values indicate easier tasks [1]. The value of difficulty 1 indicates a very easy task and a task with the value of difficulty

0 indicates a very difficult task. The value of difficulty depends on the tasks and participants. It can be limited by the presentation on the screen, the number of attempts, etc. [19]. The value of task difficulty is calculated for all the participants who took part in the Bebras contest during 2010–2015. The data of tasks on algorithms of participants from the 5th to 10th grade are processed. The interval of the value of difficulty is presented according to the task difficulty, provided by task developers (Fig. 1). The box-and-whisker plot is used to show the distribution of difficulty values graphically.

In Fig. 1, the ends of boxes show the value of the task difficulty outside the upper and lower quartiles. Vertical lines in the boxes show the median. Two lines outside the box show the highest and lowest value of the task difficulty. Note that, many tasks go in line with the difficulty level "easy", but tasks of the "medium" and "hard" level are too difficult for students. The most part of tasks has a high value of difficulty. The tasks provided for participants in 2014–2015 have the lowest value of task difficulty (for example, "easy" task had a difficulty level with value 0.1 in 2014). The value of difficulty of tasks, that represent algorithmic thinking, varied from 0.1 to 0.7 in each grade. The lowest value is found in grades 9-10 (26.28 % of correct answers). Furthermore, the value of task difficulty was smaller than 0.5 (only 50 % participants are able to solve tasks correctly) in most algorithmic tasks in 2010 and 2012. 100 % of algorithmic tasks had the difficulty value smaller than 0.5 in 2010 and 90 % - in 2012. In general, the values of task difficulty were smaller than 0.5 in 44.44 % of tasks in 2015 and 50 % in 2012.

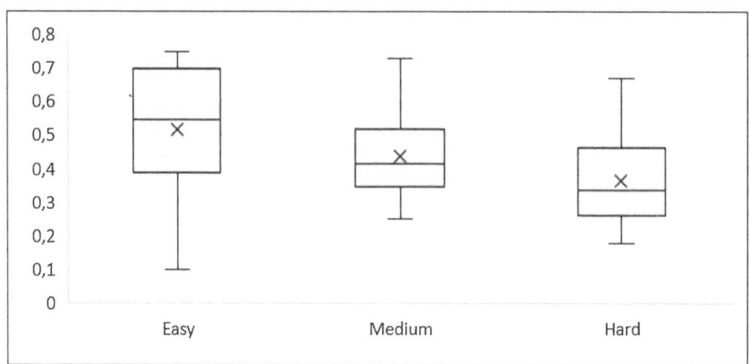

Fig. 1. Values of task difficulties distributed according to the difficulty, provided by task developers

The students' time taken for task solving correctly is related to the difficulty level. The time average was about 102 s solving easy tasks and 105 s were spent to medium tasks. The participants solved difficult tasks about 109 s. Summarizing we can say that motivation of the target group (students who solved tasks over 6 years) is not influenced by the value of task difficulty. Students' results are getting better each year (Table 4). Also, there are tasks that are very difficult to solve correctly and require more time for solving them.

Multiple choice questions are more common among the Bebras tasks. 80.4 % of such questions were focused on algorithmic skills. Only 6.6 % tasks were the tasks requiring to click something and 13.3 % of tasks consist of drag-and-drop questions. However, students spent more time for solving interactive tasks such as clicking or drag-and-drop. Students spent 181–220 s to solve interactive tasks. On the contrary, for multiple choice questions students spent twice less time, only 87–103 s.

As we have noticed, students are motivated to participate in the Bebras contest despite the difficulty of tasks.

In order to know reasons why students take part in the challenge, a deeper analysis is needed. There are several studies about task difficulties [22]. The Item Response theory (IRT) is used in most studies. IRT is usually applied to decide how students' results meet the task difficulty, estimated by task developers. We used the tasks difficulty and scores distribution to provide an overview of the interest in Bebras tasks on the long-term participation.

4   Conclusion

Students from the 3rd to 8th grades are the most active participants in the Bebras contest. Girls are interested in solving informatics tasks as well as boys. The lowest number of the participants is in the 11th and 12th grades, especially girls. The participants' number is slightly decreasing in 2014 and 2015.

Students are interested in a long-term participation. There are participants who are able to achieve the higher scores (9 participants from 137 who participated 6 years in the contest). 2 out of 9 participants who got the highest scores in a long period participation are girls.

We noticed that, students who solved less than 50 % of tasks correctly during the contest are interested in a long-term participation. They continued their participation in the challenge despite the fail on previous years. But on the other hand, the participants who have solved tasks in a long–term period are able to achieve better results than the average of group. Although, less students participate from the 11th and 12th grades, but most of them try to have better results and achieve the highest scores. Students get an experience, practice solving tasks in a long-term participation in the Bebras challenge. There is a belief that students' motivation and success are encouraged by well-balanced and interesting task content.

The value of the tasks difficulty is related to students' success. Including algorithms more difficult tasks are with difficulty "hard" (provided by tasks developers). The lowest values of tasks difficulty are noted in 2014 and 2015. Students spent more time for solving the difficult tasks.

Solving interactive tasks requires more time than solving multiple choice questions. A deeper analysis is needed to evaluate students' abilities to solve task according to the value of tasks default and type of tasks.

Acknowledgements. The research is partially supported by the Google CS4HS initiative – many thanks! Also, the authors would like to explicitly thank all members of the international Bebras challenge on informatics and computational thinking community that took part in task development and influenced in this way the outcome of this paper.

References

1. Aesaert, K., van Braak, J.: Gender and socioeconomic related differences in performance based ICT competences. Comput. Educ. **84**, 8–25 (2015)
2. Atanasova, G.E., Hristova, P.T.: Methodological aspects of the initial training of students for participation. In: Programming Contest in Proceedings of 2015 Balkan Conference on Informatics: Advances in ICT, Romania, pp. 1–9 (2015)
3. Bebras International Challenge on Informatics and Computational Thinking. http://www.bebras.org/en/facts. Accessed 30 Apr 2016
4. Bellettini, C., Lonati, V., Malchiodi, D., Monga, M., Morourgo, A., Torelli, M.: How challenging are bebras tasks? An IRT analysis based on the performance of Italian students. In: Proceedings of 2015 ACM Conference on Innovation and Technology in Computer Science Education, pp. 27–32 (2015)
5. Bucher, T.: The algorithmic imaginary: exploring the ordinary affects of Facebook algorithms. Inf. Commun. Soc., 1–15 (2016). http://dx.doi.org/10.1080/1369118X.2016.1154086
6. CSTA & ISTE: Operational definition of computational thinking for K-12 education (2011). https://csta.acm.org/Curriculum/sub/CurrFiles/CompThinkingFlyer.pdf
7. Dagienė, V., Futschek, G.: Bebras international contest on informatics and computer literacy: criteria for good tasks. In: Mittermeir, R.T., Sysło, M.M. (eds.) ISSEP 2008. LNCS, vol. 5090, pp. 19–30. Springer, Heidelberg (2008)
8. Dagiene, V., Jevsikova, T.: Reasoning on the content of informatics education for beginners. Socialiniai mokslai **78**(4), 84–90 (2012)
9. Dagienė, V., Mannila, L.A, Poranen, T., Rolandsson, L., Söderhjelm, P.: Students' performance on programming-related tasks in an informatics contest in Finland, Sweden and Lithuania. In: Proceedings of 2014 Conference on Innovation & Technology in Computer Science Education, Uppsala, Sweden, 21–25 June 2014
10. Dagienė, V., Stupurienė, G.: Bebras- a sustainable community building model for the concept based learning of informatics and computational thinking. Inform. Educ. **15**(1), 25–44 (2016)
11. Forišek, M.: The difficulty of programming contests increases. In: Hromkovič, J., Královič, R., Vahrenhold, J. (eds.) ISSEP 2010. LNCS, vol. 5941, pp. 72–85. Springer, Heidelberg (2010)
12. Futschek, G.: Algorithmic thinking: the key for understanding computer science. In: Mittermeir, R.T. (ed.) ISSEP 2006. LNCS, vol. 4226, pp. 159–168. Springer, Heidelberg (2006)
13. Yadav, A., Mayfield, Ch., Zhou, N., Hambrusch, S., Korb, J.T.: Computational thinking in elementary and secondary teacher education. ACM Trans. Comput. Educ. **14**(1), 5 (2014)
14. Kalelioğlu, F., Gülbahar, Y.: The effects of teaching programming via scratch on problem solving skills: a discussion from learners' perspective. Inform. Educ. **13**(1), 33–55 (2014)
15. Mannila, L., Dagiene, V., Demo, B., Grgurina, N., Mirolo, C., Rolandsson, L., Settle, A.: Computational thinking in K-9 education. In: Proceedings of Working Group Reports of the 2014 on Innovation & Technology in Computer Science Education Conference, ITiCSE-WGR 2014, pp. 1–29 (2014)
16. OECD: Review of Policies to Improve the Effectiveness of Resource Use in Schools (Scholl Resources review). Country Background report for Lithuania (2015). https://www.oecd.org/edu/school/Lithuania_CBR_OECD-SRR_May2015.pdf
17. Palmer, D.: Research report: a motivation view of constructivist-informed teaching. Int. J. Sci. Educ. **27**(10), 1853–1881 (2005)

18. White, R.W.: Motivation reconsidered: the concept of competence. Psychol. Rev. **66**, 297–333 (1959)
19. Peerear, J., Van Petegem, P.: Measuring integration of information and communication technology in education: an item response modelling approach. Comput. Educ. **58**, 1247–1299 (2012)
20. Perrenet, J., Groote, J.F., Kaasenbrood, E.: Exploring students' understanding of the concept of algorithm: levels of abstraction. In: Proceedings of 10th Annual SIGCSE Conference on Innovation and Technology in Computer Science Education, pp. 64–68 (2005)
21. Strain-Seymour, E., Way, W., Dolan, R.P.: Strategies and Processes for Developing Innovative Items in Large-Scale Assessments. Pearson Education, Inc., New York (2009). http://images.pearsonassessments.com/images/tmrs/StrategiesandProcessesforDeveloping InnovativeItems.pdf
22. Van der Vegt, W.: Predicting the difficulty level of Bebras task. Olymp. Inform. **7**, 132–139 (2013)
23. Vaníček, J.: Bebras informatics contest: criteria for good tasks revised. In: Gülbahar, Y., Karataş, E. (eds.) ISSEP 2014. LNCS, vol. 8730, pp. 17–28. Springer, Heidelberg (2014)

What Makes Situational Informatics Tasks Difficult?

Jiří Vaníček$^{(\boxtimes)}$

University of South Bohemia in České Budějovice,
České Budějovice, Czech Republic
vanicek@pf.jcu.cz

Abstract. Organizers of the informatics contest Bebras in many countries face the obstacle of how to state the difficulty of contest tasks and problems accurately. This is essential if tasks with good prediction of success are to be selected for the contest.

The paper discusses five different indicators of a contest task difficulty. We study which of the indicators are correlated. Index of task difficulty was defined as a combination of indicators with the aim of pinpointing the factors that have impact on a contest task difficulty. The index has been constructed on the basis of calculations from some of these indicators.

Using statistical analysis of data from Czech Beaver of Informatics contest we try to assess if some of the proposed factors can be reliable indicators of whether a task is more or less difficult. The findings show that there is a provable link between a task difficulty and the use of formalized description of the task, structuring, optimization, task assignment reading comprehension difficulty and interactivity of answers. The findings of this research study will be useful for organizers of informatics contests as well as school curricula developers.

Keywords: Bebras · Beaver of informatics · Situational informatics task · Task difficulty · Difficulty factors

1 Introduction

Organizers of informatics contests such as Beaver of informatics[1] have been consistently focusing on development of new type of informatics tasks that are not based neither on a practical activity, nor on project work and creativity. This kind of task is not "develop a program, create an algorithm, propose a solution to the problem". The new type of task is referred to as a situational task. In a situational task solvers emerge into a described situation in which they must grasp, get to understand the used concepts and terms, find an informatics principle the task is based on, solve the problem using cognitive and thinking skills and select the right answer from the offered choices. The contest tasks always have informatics background but are set in situations and environments from the contestants' lives. This allows the contestants to imagine and visualize the situation and to use their experience. No knowledge is prerequisite to

---

[1] http://www.bebras.org/, Czech national contest: http://www.ibobr.cz.

© Springer International Publishing AG 2016
A. Brodnik and F. Tort (Eds.): ISSEP 2016, LNCS 9973, pp. 90–101, 2016.
DOI: 10.1007/978-3-319-46747-4_8

solution of these tasks. This type of tasks was named after the competition "Bebras tasks".

Situational tasks can be expected to become an important part of future school informatics or computer science curricula which will focus more on development of computational thinking. Thus it is important to study the factors that affect a task difficulty. Changing some parameters of such tasks can change their difficulty or can allow their modification for different age groups. A particular informatics principle is quite likely to be comprehensible even to younger pupils if it is presented through age appropriate tasks from which those factors that make the task more difficult have been removed and factors making the task easier included.

Many authors of contest tasks have tried to analyze the contest tasks from different perspectives – selection of tasks [1], the role of pictures and illustrations [2], the process of development of a task [3], translation of tasks and their implementation in different national environments [4]. Many authors of this type of tasks have also been interested in the difficulty of tasks that are developed for the contest. The main reason is that stating a task difficulty has impact on contestants' performance and success in the contest test. The quality of tasks is crucial for the success of a contest [5]. The contest tasks are divided into three levels of difficulty with different scores. The contestant gets more points for a more difficult task if it is solved correctly and loses more points if their answer is incorrect. The organizers therefore want to be able to predict accurately which of the tasks will be seen as hard and which will be taken for easy in the contest. There are Slovak [6, 7] and other [8, 9] studies focusing on a contest task difficulty presenting qualitative analysis of individual tasks with respect to their difficulties available.

The above listed reasons made us start research into what affects a task difficulty, what criteria of a task difficulty already exist but also what factors can be used to assess a task difficulty.

2   How to Set the Test Item Difficulty

Analysis of existing literature focusing on contest task difficulty and analyses of contest systems in different states show that a task difficulty is understood as the contestants' success when solving the tasks. It is very often expressed as correct answer ratio. However, data from contests provide more information that can be used to state a task difficulty. For example if a significant number of contestants refuse to answer a particular item, the task can be assessed as very difficult because it put the contestants off attempting to solve it. A problem difficulty is thus to a certain extent also the contestants' subjective attitude, not only the objective measured number of correct answers.

Pozdnyakov and his team [9] are aware of this fact and define two parameters of difficulty. Сложность (translated as Complexity – probably good translation is Complicacy) is an objective factor and it is calculated as the ratio of participants who gave a correct answer among those who decided to solve the task. Трудность (translated as difficulty) is understood as a subjective factor, i.e. how a participant interprets a task, and is expressed as a proportion of participants who chose the "no answer" option in a particular task. Both these factors should be taken into account [9]. These parameters may provide interesting findings but do not help us define one unique index of difficulty that would allow to assess, classify and place tasks into a contest test.

Van der Vegt presents an interesting way of predicting a test task difficulty based on the use of a questionnaire for difficulty level estimation. The questions focus on two areas: The question answering process and the size of the problem [8].

3   What Affects a Task Difficulty

A difficulty of a task is affected by a number of factors that can be divided into the so called general and subject specific factors. General (formal) factors are such that can be come across in knowledge tests in general and are not specific for a particular discipline. These are for example:

- length of the text
- demands on reading comprehension (how attentively the text must be read)
- task formulation (e.g. is the posed question negative?)
- formulation of answer (e.g. presence of distractors or confusing answers, tricks)
- use of an explanatory picture
- use of an illustrative example

Subject specific factors are tied to informatics and are specific for an informatics contest. A task difficulty can then be affected e.g. by:

- area (e.g. whether optimization tasks are more difficult than algorithm tasks)
- way of solution (competences needed for the solution of the task)
- situation (to what extent the situation is close to the contestant's life experience)
- presence of formal description (code, programme, formula)
- expertness of task assignment (to what extent it uses technical terminology whose insufficient knowledge may result in failing to understand the task itself)

General factors of task difficulty in Bebras contest are discussed by Gujberová [10] who draws attention to cognitive aspects of tasks. In her study she has proven an impact of formal aspects on overall task difficulty. In this respect Pohl provides recommendations on a task formal aspects resulting in higher quality of the task, e.g. short sentences, a one-to-one relationship between words and objects, appropriate analogies [3]. Some factors have different consequences in works of different authors. E.g. Pozdnyakov states that "among numerous ways of assessment of difficulty of a text, the most straightforward is the length of the statement" [9, p. 34]. In contrast results of Gujberová's research suggest no correlation between the length of the assignment and task difficulty [10].

Some subject specific factors were studied by Křížová [11], namely the area, presence of formal description and way of solution (needed competences). These comparisons have only shown a significant correlation between presence of formal description and task difficulty.

4 Method

In the first stage of our research we focused on defining the index of a test task difficulty. At the second stage we were looking for factors that (in our opinion and in literature) have impact on a task difficulty and at the same time can be identified in assignments reliably.

The basic set of analyzed tasks was the set of all tasks from Czech national rounds of Bebras contests from 2012–2015, which is 283 contest tasks in 5 age categories for pupils from 9 to 19. The smallest number of contestants solving one task was 2492, the average number of solvers per one task was 7198.

The reason why we excluded tasks from foreign contests was linguistic – it is very difficult to analyze tasks posed in foreign languages (e.g. how to assess demands on reading comprehension from the contestant's point of view). Another reason is that different countries organize the contest differently and the systems do not provide comparable data for stating which tasks were solved successfully and which quantitative factors might have had impact on success rates (e.g. data about time needed for solving the task) (Fig. 1).

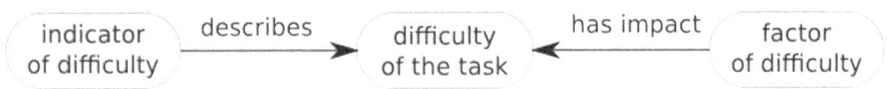

Fig. 1. Relationship between indicators and factors of task difficulty

4.1 Selection of Indicators for Stating Difficulty Index

We determined five indicators that describe and allow prediction of difficulty of Beaver of Informatics contest tasks. These indicators were compared. We explored to what extent they can be regarded as relevant. These indicators are:

- Contestants' success rate – what percentage of contestants gave a correct answer (in case of multiple choice task chose the correct alternative).
- No answer – the proportion of contestants who skipped this task, chose not to answer it.
- Authors' opinion – how the authors of the tasks or test (i.e. pedagogical experts) define the task difficulty (the tasks are classified as easy, middle and hard; a test in each category includes the same number of tasks of each difficulty).
- Solving time – how long it took the contestants to answer the task.
- Contestants' opinions – how many respondents (contestants) marked the particular task as the most difficult in a questionnaire filled in immediately after the contest.

Each of these indicators has its limits. Sometimes a different interpretation than interpretation pointing at a task difficulty may be possible. Thus the data these indicators provide cannot always be perceived as absolutely reliable.

- Contestants' success rate might be affected by cheating, by helping each other out and whispering answers. Analysis of cheating in Czech contest shows that this objection is far from purely hypothetical [12].
- No answer may also be caused by a situation when the contestants' do not have enough time to complete the test and thus do not get to some of the tasks.
- In case of tasks adopted from other countries, opinions of authors' on task difficulty may be affected by different distribution of contestants to age categories.
- Solving time is given as the difference between the time of subjecting the answer to the previous and the currently solved task. However, this may not be the time in which the contestant was really solving the given task.
- The follow-up questionnaire is voluntary, which means data from all contestants are not available.

4.2   Comparison of Indicators to State a Task Difficulty

The above listed indicators are complementary to each other and if they are combined they can determine a task difficulty more precisely and more complexly. What we tried to do was to take into account all these indicators to define the coefficient of mean difficulty. We tried to find out how individual indicators differ from each other, from the mean, what their variance is and which of the indicators best corresponds to the others.

The sources of data for this part were reduced to the results of the 2012 and 2013 national rounds of the Beaver of Informatics in upper secondary school categories. These provided the data for the first for above listed indicators. The data came from 18653 evaluated contestants in 60 tasks. The reason for this reduction of the set of data was that contestants' comments on the task difficulty given after the contest were available only for this set of tasks. These comments were from 1414 respondents of the questionnaire in which the selected the most difficult test task in the category.

For all indicators, the order of tasks for a given age category was stated as follows:

- Success rate: correct answer ratio.
- No answer: the most difficult task is skipped by the greatest number of contestants.
- Authors' opinion: the order of tasks of a given difficulty (easy, middle, hard).
- Solving time: the most difficult problem took most time to solve.
- Contestants' opinion: the task was selected as most difficult by contestants.

To be able to assess which of these indicators has most impact on a task difficulty, we studied the distance of individual indicators from the mean value of order, to what extent, if compared in pairs, the indicators provide similar results, and in how many tasks the particular indicators differ extremely from other indicators.

4.3   Search for Factors with Impact on a Task Difficulty

Based on an analysis of task text assignments we were looking for those properties of a task that are easy to detect in tasks across age categories and that are present in a significant number of tasks. For this reason some factors that looked very promising at first had to be dropped. The greatest risk especially in case of qualitative factors is the subjectivity in deciding whether a factor is present in a task or not. Full set of tasks from 2012–2015 was used for this part of research.

In the end we defined a total of 20 possible factors that were grouped according to how they can affect higher/lower difficulty of a task type: topic, way of answering, type of interactivity, other general factors and subject specific factors.

Difficulty Given by the Task Topic. The thematic area or topic of a contest task is given according to Dagienė and Futschek [13]. This is used by International Bebras Committee for categorization of proposed tasks in national contests: algorithmization, understanding information and its representation, understanding structures and problem solving. We wanted to find out whether any of these 4 categories could be factors of difficulty, i.e. whether the topic itself can determine task difficulty.

Difficulty Given by the Way of Answering. The contest test allows three types of answers:

- Multiple-choice – selection of one out of four possible answers. It limits the choice of answers, it involves distractors – trick choices and allows the solving strategy "going through all choices and comparing them".
- Textbox – entering text into a textbox (most often using a number or a word from the assignment). This form allows more variety than multiple-choice but it is sensitive to syntactical errors.
- Interactive – most often by manipulating objects on desktop, by moving them, putting them in a different order, changing their appearance by clicking. Some of the tasks were programmed as games or allowed keyboard input.

Type of Interactivity. Since we anticipated that this research will verify popularity of interactive assignments and as there are several variants of interactive solutions, interactive tasks were divided into subcategories and different factors were linked to these categories:

- Drag and drop – moving objects on desktop e.g. to the right order, making pairs
- Click – tagging objects by clicking (appearance of objects changes)
- Text – controlling interactivity by writing a text, e.g. by writing a programme code
- Game – more complex control described by rules, often resembles control of a game (puzzle, maze).

Other General Factors. General factors are those that can be come across in test tasks regardless of the discipline they come from. These are:

- length of the text – number of signs in the assignment
- demands on reading comprehension – how difficult it is to read the text and how much attention is needed (long sentences, repetition of similar words, precisely described situations). This difficulty may not necessarily be related to scientific demands of the text.
- illustrative picture – a picture that illustrates the situation, explains a concept

- example – a concrete example that illustrates the rules described in the assignment and which allows to check understanding of the assignment
- negative question – question that is formulated as e.g. "Who is not?" instead of "Who is?". Children may miss the negation in the question and answer it as a positive one

Subject Specific Factors. In informatics these are:

- Technical terminology – terms and concepts from informatics and computer science. If technical terminology is used, contestants' success will be affected by their expert knowledge which depends on the curriculum implemented at their school.
- Formal description – e.g. use of code, formula, excerpts from programmes, chains of seemingly disconnected signs, abridged description etc. Tasks that include a graph, table, diagram do not fall in this category, unless some code is included as well.
- Graphical structures – there is a graph, diagram, map or scheme in the task assignment. Contestants must be able to read information in graphical structures, grasp them, fill in data into them or to construct them from the given data.
- Optimization – optimization tasks can be perceived as a stand-alone thematic category, optimization does not appear in test task topics according to [13]. That is why we include it here. In some cases tasks looking for maximum or minimum are included in this category.

Hypotheses on impact of factors on a task difficulty were formulated for these general and subject specific factors. The null hypotheses were formulations of the type "The difference in the mean value of difficulty of tasks with the given factor and without this factor is zero". By comparing it to the task difficulty index we assessed whether the given factor affects success rate. As the hypothesis of equal variances in most of the studied factors was disproved, Welch's statistics that takes into account unequal variances was used. We considered impact of a factor to be proved if the null hypothesis was disproved on 95 % level of significance. If a test proved a statistical significant difference in comparison of mean values in both sets of tasks, it could be derived whether the given factor makes the task simpler or more difficult.

The only factor that was assessed differently was the **length of the text**. Here the parameter of stating the length was the numerical value of the number of signs in the task assignment. This allowed linear regression.

5   Research and Results

5.1   Defining a Task Difficulty Index

First we analyzed which of the indicators differs from the mean difficulty. In 16 out of 60 tasks we could observe that authors' opinion on a task difficulty is quite different from the other factors. What we found astounding was that it was the factor contestants' opinion that was closest to the mean difficulty. It differed least and also showed the smallest variance. On the contrary the indicator that differed most was solving time (which means that according to our method it gives least information on a task difficulty).

In the next step, pairs of indicators of a task difficulty were compared to analyze to what extent they provide similar results. The greatest difference between all ten compared pairs of indicators was in the pair authors' opinion – success rate. Indicators with most similar results were no answer – contestants' opinion. This would mean that contestants mark as most difficult those tasks they skipped in the test and did not answer. However, the criterion does not seem to be reasonable enough to use it for determining as the most difficult the task that put off most contestants. It is also quite interesting that the contestants' opinion states a task difficulty more accurately than authors' opinion.

We studied variance between different factors. In some tasks one indicator was significantly distant from the others and its omission would have resulted in a significantly smaller variance. The values that were left out by this in the greatest number of tasks was the indicator authors' opinion but surprisingly also the indicator success rate, which has so far been taken as the most dominant indicator of a task difficulty. The indicator that was excluded least often was the indicator no answer.

Based on results of the above described analysis, two most suitable indicators were used for definition of index of a task difficulty: incorrect answer ratio in answered tasks (signaling real difficulty) and no answer ratio (signaling perceived difficulty) according to Pozdnyakov [9]. These indicators described difficulty well together so their values were added. As the reason for not answering a task might be lack of time when solving the test causing that the contestant did not get to the task at all, this indicator does not give a task difficulty unequivocally and thus it is divided by two in the calculation.

Index of difficulty from absolute number of answers is expressed by the formula

$$i = (1 - n - c)/(1 - n) + n/2$$

where i – is index of a task difficulty, c – correct answers ratio, n – no answer ratio. This index was used by factors impact calculations on all of 283 tasks.

We were interested in how the two basic components of index of difficulty change with age of contestants. Graph on the left in Fig. 2 shows a huge difference in the

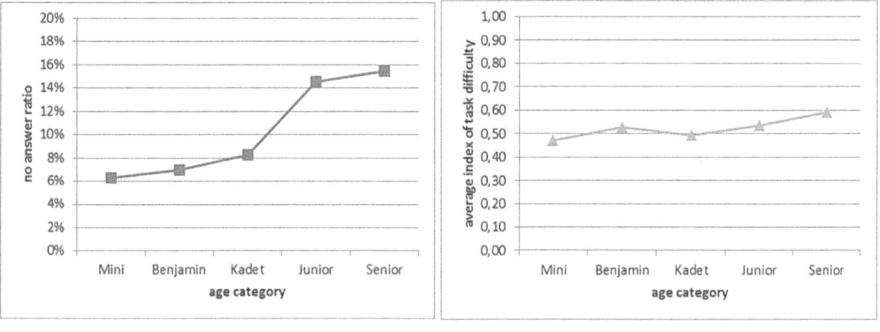

Fig. 2. Relations of unanswered tasks ratio and difficulty index to a contestant's age

number of test items with no answer at lower and upper secondary school level. The number increases abruptly at the age of 15 and almost doubles. This is in contrary to real difficulty of tasks (on the right). The reason for this might be that the contestants start to perceive tasks as difficult or that they understand the rules of the contest better and do not want to run the risk of losing points.

5.2   Factors Affecting a Task Difficulty

Proved Impact on A Task Difficulty. Statistical calculations prove impact of the following five factors:

- *Presence of formal description increases a task difficulty.* An informatics task is more difficult if it includes a code, programme, formula, chain of texts conveying some meaning. It can be inferred that the need to grasp formal description is connected to abstraction, generalization and non-trivial mental operations that the task demands from the contestant.
- *Structural tasks are more difficult.* In contrast e.g. to algorithmization or information comprehension, to looking for a structure of objects and phenomena, orientation in structures (including traditional structures as trees or containers) represent a difficulty for the contestants.
- *Presence of optimization increases a task difficulty.* Contestants face problems in the process of optimization when they are to select the best choice from a set of possible choices (which is often large).
- Demands of *reading comprehension increase a task difficulty.* It is very interesting to see that neither the length of the text, nor its difficulty have so much impact as demands on reading comprehension itself. More difficult are those tasks in which the data needed for its solution are harder to be detected in the text and remembered.
- *Interactivity of a task lowers its difficulty.* However, it depends on the type of interactivity. Of all the sub-factors into which interactivity factor was divided, only the types Drag and drop and Game actually decrease a task difficulty.

In all cases the presence of the particular factor in the task assignment brought about a significant change in the mean value of the difficulty index.

Unproven Impact on A Task Difficulty. Let us now present some potential factors whose impact on a task difficulty was not proved.

- *Presence of an illustrative picture or example* does not decrease a task difficulty. We explain this by the fact that authors decide to use such a picture or example only if they find the assignment difficult. Thus these aids are not likely to be present in easy tasks.
- The need to *work with a diagram, graph, map, scheme* has no impact on a task difficulty. This could imply that children have no problems when working with visual data or that authors of tasks are very careful when making the decision whether to use these tasks in younger categories.

- *Entering text into a textbox* (surprisingly) has no significant impact to difficulty. But we have to keep in mind that we have used this type of answer very rarely.
- A *negative question* has no impact on a task difficulty. This might be caused by the fact that this kind of question is always alerted to by its format or other warning to make sure the contestants do not fail to notice it.
- *Length of a text has no significant impact on real difficulty* (Fig. 3). The result of linear regression shows that no significant impact of the length of a text on the actual difficulty of a task (given by index of difficulty) was proved. The graph shows that in the set of 283 tasks only weak or moderate dependence of difficulty on length of the text was proved, coefficient of determination R^2 in linear regression model was only 0.11 and Pearson's correlation coefficient 0.34.
- *The length of the text has no significant impact on perceived difficulty.* Significant impact of the length of the text on perceived difficult was not proved (given by no answer ratio). Pearson's correlation coefficient 0.55 indicates moderate dependence. So our results agree with [11] and disagree with [9] in this case.

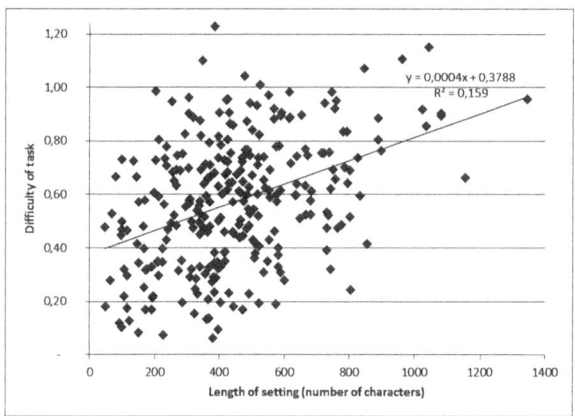

Fig. 3. A low dependence of a task difficulty index on the length of assignment text.

Perceived Task Difficulty in Contrast to Its Real Difficulty. In 5 out of 20 assessed factors we could observe a difference between actual and perceived test task difficulty (given by no answer ratio). This means that the impact of some factors on real task difficulty did not affect the real difficulty statistically but affected the perceived difficulty (or vice versa).

Tasks of type problem solving (strategies, logical tasks) were perceived as easier by the contestants but the ratio of correct answer showed no difference. Multiple-choice tasks were perceived as more difficult but in reality they were not. Also subtype of interactive tasks with clicking on objects was perceived as more difficult. Optimization problems were not perceived as more difficult but in reality were more difficult. What we found most striking was that the presence of an illustrative picture makes the task appear more difficult to the contestants, as this type of tasks was answered by relatively fewer contestants. This might be explained by the assumption that the author decided to use this illustration out of their fear that the task was too difficult. The author seems to anticipate this difficulty thanks to some other signal. And contestants seem to perceive the task difficulty intuitively thanks to the same signal. It is this "other" signal, not the presence of an illustration or example that causes more refusals to solve the task. If the contestant decides to solve the problem in the end, the example will help them and the task will not be as difficult in the end.

6   Conclusion

Comparison of different indicators for stating difficulty of contest tasks shows that the criterion of success rate (i.e. what proportion of contestants chose the correct answer) does not always correspond to other indicators and that the indicator no answer describes a task difficulty very well. Thus a task difficulty will be more accurately predicted if we take into account not only the indicator success rate but also the indicator no answer. Thus our recommendation is to use the rule based on both of these indicators.

Presence of formalized description, structuring, optimization and demands of assignment reading comprehension are also substantial factors making a task more difficult. Interactivity of answering appears to be a significant factor making the task less difficult. On the other hand factors as the length of the text, use of technical terminology, algorithmization, diagrams, negative questions, illustrative pictures or examples did not prove to affect the difficulty of an informatics task.

It must be stressed at this point that the discovered statistical dependencies do not automatically mean there is some causality. We do not claim that the presence of any of these indicators and factors is responsible for a task difficulty. For example the cause of difficulty of structuring and formal description may be the higher level of necessary abstraction. However, this is hard to detect. We had to look for factors that can be detected in a task assignment more or less accurately and use such for determination of a task difficulty. The real difficulty of a task will always have to be detected only experimentally, i.e. when it is actually solved in the contest.

Results of this research study cast some light on the area of predicting difficulty of situational informatics tasks and at the same time indicate in which direction to continue in research: to look for links between individual factors, to study these factors in different age and gender groups, in a child's development, compare these factors across countries or school subjects.

Accurate determination of the difficulty of test tasks for an informatics contest will be of benefit when constructing contest tests, for development of new tasks designed for a particular age group or level of difficulty. If some elements of a task prove to contribute to its difficulty, it will be possible to modify the tasks and tune the test. Also findings on sources of difficulties of informatics tasks may contribute to development of higher quality informatics curricula.

Acknowledgment.   The research was supported by the project GAJU 121/2016/S.

References

1. Dagienė, V.: What kinds of tasks are good for contests? In: 6th International Conference on Creativity in Mathematics Education and the Education of Gifted Students, Riga, Latvia, pp. 62–65 (2011). ISBN 9789984453606
2. Tomcsányiová, M., Kabátová, M.: Categorization of pictures in tasks of the Bebras contest. In: Diethelm, I., Mittermeir, R.T. (eds.) ISSEP 2013. LNCS, vol. 7780, pp. 184–195. Springer, Heidelberg (2013). ISBN 978-3-642-36616-1
3. Pohl, W., Hein, H.-W.: Aspects of quality in the presentation of informatics challenge tasks. In: Jekovec, M. (ed.) The Proceedings of International Conference on Informatics in Schools: Situation, Evolution and Perspectives — ISSEP 2015, pp. 21–32. Založba FRI, Ljubljana (2015)
4. Tomcsányi, P., Vaníček, J.: International comparison of problems from an informatic contest. In: ICTE 2009: Information and Communication Technology in Education 2009, pp. 219–221. University of Ostrava, Ostrava (CZ) (2009). ISBN 978-80-7368-459-4
5. Dagienė, V., Stupurienė, G.: Bebras – a sustainable community building model for the concept based learning of informatics and computational thinking. Inform. Educ. 15(1), 25–44 (2016)
6. Tomcsányi, P.: Náročnosť úloh v súťaži Informatický bobor (Difficulty of tasks in Informatický bobr contest. In: Konferencia DidInfo 2009. Univerzita Mateja Bela, Banská Bystrica (SK) (2009). ISBN 978-80-8083-720-4
7. Tomcsányiová, M., Tomcsányi, P.: Analýza riešení úloh súťaže iBobor v školskom roku 2013/14. (Analysis of solutions in iBobor contest tasks in 2012/13 year). Konference DidactIG 2014. Technical University, Liberec (CZ) (2014)
8. van der Vegt, W.: Predicting the difficulty level of a Bebras task. Olympiads Inform. 7, 132–139 (2013)
9. Yagunova, E., Podznyakov, S., Ryzhova, N., Razumovskaia, E., Korovkin, N.: Tasks classification and age differences in task perception. Case study of international on-line competition "Beaver". In: Jekovec, M. (ed.) The Proceedings of International Conference on Informatics in Schools: Situation, Evolution and Perspectives — ISSEP 2015, pp. 33–43. Založba FRI, Ljubljana (2015)
10. Gujberová, M.: Výber úloh do informatickej súťaže iBobor (Collection tasks for informatics contest iBobor. In: Konference DidactIG 2014. Technical Universoty, Liberec (CZ) (2014)
11. Vaníček, J., Křížová, M.: Kritéria obtížnosti testových otázek v informatické soutěži (criteria of informatics contest tasks difficulty). In: Lovászová, G. (ed.) DidInfo 2014, pp. 191–199. Univerzita Mateja Béla, Banská Bystrica (SK) (2014)
12. Šimandl, V.: Odhalování podvádění v online soutěžích (detecting of cheating in online competitions). J. Technol. Inf. Educ. 6(2), 114–121 (2014). ISSN 1803-537X
13. Dagienė, V., Futschek, G.: Bebras international contest on informatics and computer literacy: criteria for good tasks. In: Mittermeir, R.T., Sysło, M.M. (eds.) ISSEP 2008. LNCS, vol. 5090, pp. 19–30. Springer, Heidelberg (2008)

Best-Practice Papers
and Country Reports

A New Informatics Curriculum for Secondary Education in The Netherlands

Erik Barendsen[1]([⊠]), Nataša Grgurina[2], and Jos Tolboom[3]

[1] Radboud University and Open University, Nijmegen, The Netherlands
e.barendsen@cs.ru.nl
[2] University of Groningen, Groningen, The Netherlands
n.grgurina@rug.nl
[3] SLO Institute for Curriculum Development, Enschede, The Netherlands
j.tolboom@slo.nl

Abstract. In The Netherlands, the current informatics curriculum for upper secondary education was introduced in 1998 and only slightly modified in 2007. Meanwhile, both the scientific discipline and its impact on society have developed substantially. For this main reason, a curriculum reform has been carried out which has led to a new curriculum specifying the intended learning outcomes. This country report specifies the educational context in which the reform takes place. Moreover, it decribes the reform process from various perspectives, highlights and explains the underlying design principles that guided the development of the new curriculum, and presents its main results.

Keywords: Informatics · Curriculum · Secondary education · Reform

1 Introduction

This country report focuses on the ongoing curriculum reform for the elective informatics subject in upper secondary education in The Netherlands. The intended learning outcomes have recently been specified by a curriculum commitee. The current state of affairs – after completion of the learning outcomes, but before the implementation of the curriculum in schools – seems a good occasion to discuss the background, design principles and content of the new curriculum.

Since 2010, there has been a growing concern about informatics education among European and American teachers, scientists and industry [10,13]. Remarkably, in recent years, the call for reform of informatics education could increasingly be heard *outside* the informatics community [1,9,14,17]. By now, several countries have already innovated their informatics education.

In the Dutch educational system, informatics is an elective subject in upper secondary education. Its curriculum was established in 1998, and was adjusted only slightly in 2007. The subject has been evaluated several times, last time in 2007 [18], but this has not led to major modifications. This is remarkable, as informatics curricula in higher education have been adapted several times since 1998.

A. Brodnik and F. Tort (Eds.): ISSEP 2016, LNCS 9973, pp. 105–117, 2016.
DOI: 10.1007/978-3-319-46747-4_9

Reference curricula have been periodically updated, accommodating new themes [2]. Moreover, all science subjects in Dutch secondary education have been thoroughly revised in the past few years.

Triggered by concerns expressed by the Dutch academy of sciences [14], the Ministry of Education ordered a study [20] into the teaching practice of the subject, see also [4]. The researchers conducted a literature review, teacher interviews, a survey and expert consultations. The questionnaire was completed by a representative part for the teacher population.

The study confirmed that the situation was alarming. This led the Ministry of Education to decide in favour of a curriculum reform for informatics in the upper level of upper secondary education. A curriculum committee of 9 members was installed in September 2014, in which teachers, computing education specialists, experts from universities and universities of applied sciences, and curriculum and assessment specialists were represented. The curriculum was conceived in 2015 and the advisory report was formally presented in March 2016. The ministry is expected to formally adopt the curriculum by the summer of 2016. The curriculum will come into effect in schools by August 2019.

This paper is structured as follows. We start with an introduction to the Dutch educational context in Sect. 2. In Sect. 3, we explore some challenges for the new curriculum and explain the design principles underlying the curriculum and its description. An outline and some characteristic examples of the contents of the curriculum are described in Sect. 4.

2   Educational Context in The Netherlands

2.1   The Dutch Educational System

In the Netherlands, students complete elementary school at the age of twelve. The Dutch system offers three main types of secondary education, two of which offer informatics.

The HAVO type of school (senior general secondary education, in Dutch: *hoger algemeen vormend onderwijs*) spans five years (grades 7 through 11) and prepares students for higher professional education, while the VWO type of school (pre-university education, in Dutch: *voorbereidend wetenschappelijk onderwijs*) spans six years (grades 7 through 12) and is geared towards further education at a university.

The final assessment of secondary education consists of two parts: national exams and school exams. Nearly all subjects have both type of exams — except informatics and a few others, which only have a school exam. The Dutch government decides on the curriculum, which prescribes the learning objectives and the way they are assessed (through a school exam only or through both a school and a national exam).

Dutch curricula are formulated as a collection of learning objectives, i.e., the knowledge, insight, skills and professional attitude that a student should have acquired upon completion of the subject. It is common to formulate a global, considerably abstract description for each of the intended learning outcomes. This approach is in line with the Dutch principle of letting the government only prescribe *what* is learned; the *'how'* is left to teachers, schools and textbook authors. For example, the time dedicated to teaching a particular learning objective is for the teachers to decide.

In HAVO and VWO almost all students have the same curriculum in grades 7 through 9 with respect to most of the subjects. From grade 10 on, they choose one of the following four *tracks*: *Culture and Society*, *Economics and Societys*, *Nature and Health*, and *Nature and Technology*. Informatics can be chosen as an elective subject within all tracks.

2.2   Informatics in Dutch Upper Secondary Education

In 1998, the subject of Informatics was designed to be well within the capabilities of all students, regardless of which track they choose. This resulted in a subject of a multidisciplinary nature. Furthermore, since Informatics was not a prerequisite for any subsequent study at the university or college level, it was felt that there was no need for a national exam: the subject would be assessed through a school exam.

The student workload of the secondary Informatics curriculum is 320 study hours for HAVO and 440 hours for VWO. The first curriculum drew its inspiration from the 1994 UNESCO/IFIP curriculum [19].

The learning objectives in Dutch curricula are grouped into so-called *domains*, each of which is subdivided into one or more *subdomains*. Each subdomain is then specified by a specific learning goal. The 1998 curriculum contained 53 learning goals described in great detail. In the 2007 revision, the curriculum was brought down to 18 learning goals described in general terms. In Table 1 we present the domains and subdomains of the 2007 revision. For a detailed description of this curriculum and its implementation, see [12].

Table 1. Domains and subdomains in informatics, 1998 curriculum after 2007 revision

Domain	Subdomains
A: Informatics in perspective	Science and technology, society, study and career, the individual
B: Terminology and skills	Data representation in a computer, hardware, software, organizations
C: Systems and their structures	Communication and networks, operating systems, systems in practices, development of information systems, information flow, information analysis, relational databases, human-computer interaction, system development lifecycle
D: Usage in a context	

3 Design Principles

In this section, we will discuss some of the challenges of the curriculum innovation, each together with the chosen solution in terms of design principles.

Towards a Sustainable Curriculum Specification. The current curriculum description is considered outdated. Indeed, it stems from a time when the world wide web was used very little, and corporate automation was the main application of computers. When asked for their opinion about the curriculum, many teachers indicated that they were missing modern applications of computer science such as games and mobile devices, apart from recent fundamental themes such as security and data science [4,20].

For the curriculum committee, one of the main goals of the new curriculum was to ensure relevancy in the long term. Observing the rapid development ('revolutions') of applications and the more 'evolutionary' development of fundamental concepts, the committee decided to follow a so-called *concept-context* approach.

This approach, often referred to as context-based education, is mainly known as a *pedagogical* principle and has been applied in several science subjects, e.g., [3,8]. In this approach, science content is connected with social, professional and scientific contexts [11] which appeal to a broad group of students. The concept-context approach aims, among other things, to change students' attitudes towards science and make science more 'meaningful' and more 'relevant' to students [6,15].

The curriculum committee applied the concept-context idea as a *description mechanism* by separating fundamental concepts from the more volatile contexts (i.e., application domains and situations). In the curriculum description, the concepts have been specified concretely, but the contexts only in a generic way. In line with the concept context approach in the science subjects, the commitee distinguishes between: social contexts (including environmental contexts), professional contexts, and scientific contexts. Other important contexts for informatics are other school subjects (that apply informatics).

In order to help teachers and textbook authors to design appropriate contexts by instantiating the generic context descriptions, the institute for curriculum development (SLO) will develop guidelines for the implementation of the curriculum in classroom teaching and assessment.

Dealing with Diversity. Dutch students in informatics vary greatly with respect to interest and computing experience [20]. A complicating factor is that the subject is not only an elective within the *Nature and Technology* track, but should cater for students of *all* tracks.

Differentation is facilitated by two mechanisms. First, the new curriculum is divided into a *core* curriculum, intended for all students, and a collection of *elective* themes. The core curriculum has been composed in such a way that all students are at least able to communicate with peers who have specialised through a particular elective theme. This opens possibilities for working in project teams consisting of students with different expertise. A second mechanism is the concept-context approach described above. Since the curriculum is parametric with respect to contexts, these contexts can be differentiated according to the chosen tracks or specific personal interests of students. This can be done, e.g., by distinguishing social, economic, cultural and technical contexts.

Differentiation Between School Types. The curricula for HAVO and VWO will differ at least in extent: the vwo students spend more time on the subject than their HAVO counterparts. This is taken into account via the number of elective themes to be chosen: two for HAVO students and four for VWO students.

Making a distinction with respect to *attainment level* of the learning goals turned out to be difficult. Dutch science curricula for HAVO and VWO often differ with respect to complexity and abstraction levels. The curriculum committee has included distinct VWO and HAVO variants of the goals in a few cases where this seemed logical and obvious. In general, however, characterizing the difference between HAVO and VWO students in informatics turned out to be difficult. At a focus group session, the teachers agreed that each of them could think of a student in their HAVO classes that would also excel in the corresponding VWO class. In the case of informatics, the difference between individual HAVO and VWO students is often smaller than the difference between students from the *Culture and Society* track and students from the *Nature and Technology* track. Therefore the curriculum committee decided to refrain from further differentiation on the formal curriculum level.

Informatics and Engineering. Informatics is seen by many as a *constructive discipline*: a subject area where *creating* things (mostly digital artefacts) is the key element [7]. In this epistemic view, informatics as a scientific discipline supplies the conceptual and procedural knowledge about such artefacts and the creation process.

The 'creation' perspective is an attractive starting point for the subject. It offers a recognisable and dynamic look, and provides a nice contrast with other, more analytical, science subjects. Moreover, creating digital products does not require any complicated physical tools or materials: in most cases a computer will suffice. This means that students are not limited to designing a product (as is mostly the case in other science subjects), but will actually be able to build it. Finally, the teachers consulted during the curriculum reform often pointed to the *fun factor*: creating digital products can be fun and engaging.

Therefore, the committee has decided to position 'design and development' as a central skill in the new curriculum.

Balancing Guidance and Freedom. One of the goals of the Curriculum Committee was to find a balance between providing guidance and offering freedom. According to the informatics teachers, the previous curriculum was too

general and therefore provided too little guidance [20]. Therefore, the committee
has drafted more comprehensive learning goals. The core curriculum contains 30
learning goals, of which 13 are in the Skills domain. The remaining 34 learing
goals describe the elective themes that offer schools the opportunity to shape
the informatics education according to their own preference. The committee feels
that this creates the intended balance between providing guidance and offering
freedom.

4   The New Curriculum

In this section we will depict the overall structure of the curriculum. Then we will
describe the domains of the core curriculum and give some examples of elective
themes. The complete curriculum can be found in [5].

4.1   Curriculum Structure

The *core curriculum* consists of a *skills* domain and five *knowledge* domains. The
skills domain (A) addresses both informatics specific skills and general scientific
and technical skills. The latter are roughly the same as for science subjects such
as chemistry and physics. The elective themes span domains G–R. See Table 2.
The HAVO students choose two elective themes, whereas the VWO students select
four.

Table 2. Domains of the informatics curriculum

Core curriculum	Elective themes
Domain A Skills	Domain G Algorithmics, computability and logic
Domain B Foundations	Domain H Databases
Domain C Information	Domain I Cognitive computing
Domain D Programming	Domain J Programming paradigms
Domain E Architecture	Domain K Computer architecture
Domain F Interaction	Domain L Networks
	Domain M Physical computing
	Domain N Security
	Domain O Usability
	Domain P User experience
	Domain Q Social and individual impact of informatics
	Domain R Computational science

The following three sections will offer more explanation about the elements
of the curriculum: first, the Skills domain, then the knowledge domains of the
core curriculum, and finally the knowledge domains for the elective themes.

Grey boxes are used to display text segments from the curriculum description. The curriculum often refers to *digital artefacts*. This collective term is used to indicate products that have been designed and/or developed and/or produced based on informatics knowledge: programs, computer systems, interfaces, etc.

4.2   Domain A: Skills

Three types of skills are considered essential and characteristic for the subject of informatics: (1) designing and developing, (2) using informatics as perspective, and (3) cooperation and interdisciplinarity. We will describe these central skills below in the form of their subdomains. Other skills include the subdomains *Ethical conduct*, *Using informatics tools*, and *Working in contexts*. Moreover, for symmetry and completeness, the committee has included general skills and scientific skills from other science curricula.

Subdomain: Designing and Developing. This subdomain concerns development of digital artefacts. In the new curriculum this is considered as more than just a technical affair. Instead, contexts and users are the starting points. These contexts offer the opportunity to connect to the interests and the tracks of the students, allowing designing and developing to become creative challenges, more than just solving ready-made tasks to write a computer program, cf. [16].

Designing and developing
The candidate is able
  - to spot opportunities for the application of digital artefacts in a certain context;
  - to translate these opportunities into a design and development objective, taking the technical, environmental and human factors into account;
  - to specify the wishes and requirements, and to assess their feasibility;
  - to design a digital artefact;
  - to weigh the options in the design of a digital artefact by means of research and experimentation;
  - to implement a digital artefact;
  - to evaluate the quality of digital artefacts,
and to combine the above skills in the development of digital artefacts.

The knowledge domains (B–F and G–R) will supply knowledge and tools to execute developmental activities, such as programming, but also social impact and quality aspects like user interaction, correctness, efficiency and security.

The formulation of the above learning goal allows to vary the degrees of freedom with respect to the students' formulation of the development objective. The committee regards this as a valuable option for differentiation. For instance, an nontrivial open situation, where the formulation of a suitable objective requires consideration and research, will likely suit VWO students, while a concrete development objective seems to be a typical starting point for the HAVO students. There is a lot of room for variation between these two extremes; this depends on apects such as the complexity of the context. The committee has decided against formulating a generic distinction between HAVO and VWO in this subdomain.

Subdomain: Using Informatics as a Perspective. Informatics has an analytical element, in addition to the creative perspective: knowledge in this area allows students to explain certain phenomena in their daily lives and in society. A student who completes this subject will not only be able to accurately discuss the possible causes for the malfunctioning of the home Wi-Fi network, but will also be able to explain whether we should worry about our bank balance when an on-line banking site is unavailable due to a DDoS attack.

Using informatics as a perspective
In contexts, the candidate is able to
 - indicate, interpret and explain phenomena in terms of informatics;
 - recognise and interconnect informatics concepts;
 - estimate the possibilities and limitations of digital artefacts and reason about these in terms of informatics concepts.

This learning goal contains elements of *computational thinking* [21]. This includes *analytical skills* to formulate problems in such a way that one can use computers and other tools to help solve them, as well as *problem solving skills*, such as finding solutions in terms of algorithms and data. When 'using informatics as perspective', students use informatics concepts as spectacles to look at the world around them. This skill matches the analytical aspect of computational thinking, while our interpretation of 'designing and developing' links up with the aspect of problem solving.

Subdomain: Cooperation and Interdisciplinarity. Computer scientists rarely work alone. Creating digital products is usually a team effort: on the one hand because the products are often complex, on the other hand because the development requires expertise from various fields. Through the choice of core domains, the committee intended to enable this latter form of cooperation. For example, all students will be sufficiently informed about man-machine interaction (core curriculum, domain F) to cooperate effectively with a team member who has learned to design user interfaces (elective theme, domain O).

Software engineering has yielded various methods for structured teamwise development. The committee considers it important that students are able to use at least one of these structured approaches, but has refrained from prescribing a specific version.

Cooperation and interdisciplinarity
The candidate is able
 - to structurally cooperate in a team during the design and development of digital artefacts;
 - to cooperate with people from an application field.

4.3   Knowledge Domains of the Core Curriculum

Domain B: Foundations. This domain provides an abstract view on digital artefacts independent of any concrete implementation.

Algorithms
The candidate is able
- to develop a provisional solution for a problem into an algorithm;
- to recognise and apply standard algorithms;
- to investigate the correctness and efficiency of digital artefacts using the underlying algorithms.

Data structures
The candidate is able to compare the elegance and efficiency of abstract data structures.

Automata
The candidate is able to use finite automata for the characterisation of certain algorithms.

Grammars
The candidate is able to use grammars as tools for the description of languages.

The underlying rationale for the incorporation of automata and grammars is the following. Automata (state transition systems) often allow for an elegant and simple description or design of artefacts without directly lapsing into program code. Grammars also offer structure (and simplicity) to the descriptions of (programming) languages. These two theoretical models are intended to be used as *instruments*, rather than objects of a deep theoretical study.

Domain C: Information. This domain deals with information and concrete data. The identification and representation of data is the core of this domain, with special attention paid to the representation of numbers and media, and to the use of existing data bases.

Objectives
The candidate is able to distinguish specific purposes for information and data processing, such as searching and processing.

Identifying
The candidate is able to identify information and data in contexts, taking the purpose into account.

Representing
The candidate is able to represent data in a suitable data structure, keeping the purpose in mind; the candidate is able to compare the elegance, efficiency and implementability of various representations.

Standard representations
The candidate is able to use standard representations of numerical data and media, and is able to relate these to each other.

Structured data
The candidate is able to translate an information request into a search query for a collection of structured data.

Domain D: Programming. The Programming domain is related to both the development and the understanding and adaptation of programs. We expect students to gain skills in handling an imperative programming language. The selection of the language is left to the teachers and authors. As software is often built up from components, we use the term 'program component' instead of the classical term 'program'.

Developing
The candidate is able, for a given objective,
- to develop program components in an imperative programming language;
- to use programming language constructs that support this abstraction;
- to structure a program component in such a way that they can be easily understood and evaluated by others.

Inspecting and adapting
The candidate is able
- to explain the structure and functioning of certain program components;
- to adapt such program components based on evaluation or changing requirements.

Domain E: Architecture. The core curriculum focusses on the structure and operation of digital artefacts. Architectural elements are a good stepping stone for the introduction of security aspects, insofar as they are related to technical risks and measures. The human factors related to security are part of the Interaction core domain.

Decomposition
The candidate is able to explain the structure and functioning of digital artefacts through architectural elements, i.e., in terms of the physical, logical and application layer levels, and in terms of the components in these layers together with their interaction.

Security
The candidate is able to name some security threats and common technical measures, and relate these to architectural elements.

Domain F: Interaction. This domain is related to the link between informatics and the environment. The committee has added three perspectives to the core curriculum: the interaction between digital artefact and user (usability), the impact of informatics on society, and the impact of informatics on the individual (privacy in particular). These three perspectives offer good starting points for the socio-technical part of security, as an addition to the more technical aspects that are part of the Architecture domain.

Usability
The candidate is able
- to evaluate user interfaces for digital artefacts using heuristics;
- to apply principles of good design when designing and developing interfaces of digital artefacts.

Social aspects
The candidate is able to recognise the impact of digital artefacts on social interaction and individual privacy, and is able to put this in a historical perspective.

Privacy
The candidate is able to reason about the consequences for personal freedom given the increasing possibilities of digital artefacts.
Security
The candidate is able to recognise security threats and common socio-technical measures, and relate these to social and human factors.

4.4   Elective Themes

The elective themes offer students the opportunitys to deepen (e.g., Algorithms, computability and logic) or broaden (e.g., Cognitive computing) their informatics skills and knowledge. Below we give three examples of elective themes.

Domain J: Elective Theme Programming Paradigms. The core curriculum includes learning to program in at least one imperative programming language. In this elective theme, the students have the possibility to broaden their programming repertoire with, e.g., functional, logical, or object oriented paradigms.

Alternative programming paradigm
The candidate is able to describe the characteristics of at least one additional programming paradigm, and is able to develop and evaluate programs according to that paradigm.
Choosing a programming paradigm
The candidate is able to assess the suitability of programming paradigms for the solution of a particular problem.

Domain P: Elective Theme User Experience. This optional theme deals with digital artefacts that require a user interaction design that is more extensive than the design of a user interface. This is more advanced than the scope of the elective theme O, *Usability*. Possible examples are interactive artefacts such as games, and applications that exploit the user interaction to increase the user engagement, such as gamification of existing applications. User experience may not just be influenced by the direct interaction, but also by other modalities, such as imaging (animation, video) and sound (music).

Analysis
The candidate is able to explain the relationship between the design choices for an interactive digital artefact and the expected cognitive, behavioural and affective changes or experiences.
Design
The candidate is able to design the user interaction for a digital artefact, justify the design decisions, and implement it for a simple application.

Domain R: Elective Theme Computational Science. This domain is an extension of the sub-domain Modelling of the Skills domain A in the core curriculum. As many of our students are going to pursue scientific and engineering careers outside of informatics, they need to be equipped with skills that allow them to formulate problems in such a way that these can be solved with the help of a computer, cf. [21].

Modelling
The candidate is able to model aspects of another scientific discipline in computational terms.
Simulating
The candidate is able to construct models and simulations, and to use these for the research of phenomena in that other science field.

References

1. Académie des Sciences: L'enseignement de l'informatique en France: Il est urgent de ne plus attendre (2013)
2. ACM/IEEE-CS Joint Task Force on Computing Curricula: Computer Science Curricula 2013. Technical report, ACM Press and IEEE Computer Society Press, December 2013
3. Apotheker, J., Bulte, A., De Kleijn, E., Van Koten, G., Meinema, H., Seller, F.: Scheikunde in de dynamiek van de toekomst. Over de ontwikkeling van scheikunde in de school van de 21e eeuw. Eindrapport van de Stuurgroep Nieuwe Scheikunde 2004–2010. SLO, Enschede (2010)
4. Barendsen, E., Fisser, P., Krüger, J., Tolboom, J.: Herziening van het Nederlandse informaticacurriculum havo-vwo. Paper presented at ORD 2014, Groningen (2014)
5. Barendsen, E., Tolboom, J.: Advies Examenprogramma Informatica vwo-havo: inhoud en invoering. SLO, Enschede (2016)
6. Bennett, J., Holman, J.: Context-based approaches to the teaching of chemistry: what are they and what are their effects? In: Gilbert, J.K., De Jong, O., Justi, R., Treagust, D.F., Van Driel, J.H. (eds.) Chemical Education: Towards Research-Based Practice, pp. 165–184. Kluwer, Dordrecht (2002)
7. Berglund, A., Lister, R.: Introductory programming and the didactic triangle. In: Proceedings of the Twelfth Australasian Conference on Computing Education, vol. 103, pp. 35–44. Australian Computer Society, Inc. (2010)
8. Boersma, K.T., Van Graft, M., Harteveld, A., De Hullu, E., De Knecht-van Eekelen, A., Mazereeuw, M.: Leerlijn biologie van 4 tot 18 jaar. Uitwerking van de concept-contextbenadering tot doelstellingen voor het biologieonderwijs. CVBO, Utrecht (2007)
9. Furber, S.: Shut Down or Restart? The Way Forward for Computing in UK Schools. The Royal Society, London (2012)
10. Gander, W., Petit, A., Berry, G., Demo, B., Vahrenhold, J., McGettrick, A., Boyle, R., Drechsler, M., Mendelson, A., Stephenson, C., Ghezzi, C., Meyer, B.: Informatics education: Europe cannot afford to miss the boat. Report of the Joint Informatics Europe & ACM Europe Working Group on Informatics Education (2013)
11. Gilbert, J.K.: On the nature of 'context' in chemical education. Int. J. Sci. Educ. **28**(9), 957–976 (2006)

12. Grgurina, N., Tolboom, J.: The first decade of informatics in Dutch high schools. Inf. Educ. **7**(1), 55–74 (2008)
13. Kaczmarczyk, L., Dopplick, R.: Preparing students for computing workforce needs in the US. ACM SIGCSE Bull. **46**(2), 8–8 (2014)
14. KNAW: Digitale geletterdheid in het voortgezet onderwijs: vaardigheden en attitudes voor de 21ste eeuw. Koninklijke Nederlandse Akademie van Wetenschappen, Amsterdam (2012)
15. Koubek, J., Schulte, C., Schulze, P., Witten, H.: Informatik im Kontext (IniK): Ein integratives Unterrichtskonzept für den Informatikunterricht. In: INFOS, pp. 268–279 (2009)
16. Romeike, R.: What's my challenge? The forgotten part of problem solving in computer science education. In: Mittermeir, R.T., Sysło, M.M. (eds.) ISSEP 2008. LNCS, vol. 5090, pp. 122–133. Springer, Heidelberg (2008)
17. Samaey, G., Van Remortel, J., Bersini, H., Bruynseraede, Y., Dekelver, J., Laender, F.D., Deschoolmeester, D., Gistelinck, P., Martens, B., Martens, L., Neven, F., Snoeck, M., Steels, L., Tempels, M., Vandenabeele, P., Vanderborght, B., Vandewalle, J., Uffelen, S.V., Vercauteren, G., Verschaffel, L., Waerniers, P., wyffels, F.: Informaticawetenschappen in het leerplichtonderwijs. Koninklijke Vlaamse Academie van België voor Wetenschappen en Kunsten, Brussel (2014)
18. Schmidt, V.: Vakdossier Informatica. SLO, Enschede (2007)
19. Tinsley, J.D., Van Weert, T.J.: Informatics for secondary education, a curriculum for schools (1994)
20. Tolboom, J., Krüger, J., Grgurina, N.: Informatica in de bovenbouw havo/vwo: Naar aantrekkelijk en actueel onderwijs in informatica. SLO, Enschede (2014)
21. Wing, J.M.: Computational thinking. Commun. ACM **49**(3), 33–35 (2006)

And Now What Do We Do with Our Schoolchildren?

G. Barbara Demo[✉]

Dipartimento di Informatica – Università di Torino,
Corso Svizzera 185, 10149 Torino, Italy
barbara@di.unito.it

Abstract. After training activities introducing computing, many teachers seem to ask themselves "And now: what do we do with our schoolchildren?". When we had to design a course for about twenty-five teachers, from twelve primary and middle schools, our first concern was trying to change this. For about twelve hours of the course meetings we introduced programming, then we discussed with attendees three frameworks of long programming activities: telling stories, creating group and class quizzes and relative answers, inventing riddles modeled by equations. The participants were asked to invent, design and implement an activity adapting to their students one of the frameworks discussed. Some were able to actually work with their pupils in schools, at least partly, during the course as we invited. This approach is of interest in other teachers training courses since the suggested activity frameworks can be inflected with contents that schoolchildren are dealing with in school. Also, the teachers are helped overcoming their apprehension in proposing a first long activity to their students because they can carry on such an experience during the training.

Keywords: Teachers training · Primary and middle school · New curricula · Questioning · Activity framework

1 Introduction

The Scuola2.0 project was promoted in 2014 by the Municipality of Torino (Italy) for improving digital competencies in primary and middle schools. One of the components of the initiative was the request to organize a pilot course during the school year 2015–16 for about twenty-five in-service teachers. From our previous retraining experiences we knew to have several constraints: most of the in-service teachers have little time for retraining and, even younger ones, know little about computing. Almost all only have experienced digital literacy activities, i.e. they have been using the web and computer tools specific for an educational purpose [5]. Other peculiar constraints of the project were that only few of the teachers were volunteers and that we had to cut into two phases the meetings having half of them during autumn 2015 and the other half from beginning March 2016 to the end of April. We ended up organizing ten meetings of three hours each.

© Springer International Publishing AG 2016
A. Brodnik and F. Tort (Eds.): ISSEP 2016, LNCS 9973, pp. 118–129, 2016.
DOI: 10.1007/978-3-319-46747-4_10

In our country there is no mandatory curriculum for informatics in education neither commonly accepted suggestion when optional activities are possible. There are several formal and informal proposals going from educational robotics to programming (using different languages), from *CS Unplugged* to *coderdojo* and *fablab* type of activities, naming just some of the most popular ones. Also when deciding the contents of retraining projects we need to consider recent criticisms toward the digital presence in education till nowadays coming from different sources, for example OECD [8].

After having for years experienced the above mentioned approaches, we support introducing basic principles of computing by programming as a contribution to defining new digital curricula for schools. With our proposals we are then in line with the suggestions by Schulte in [11], Ben Ari in [1] and with those coming from the English National curriculum for primary school where we read:

> The role of programming in computer science is similar to that of practical work in the other sciences: it provides motivation, and a context within which ideas are brought to life [2].

Moreover, our project concerns k-8 education: thus a practical programming work complies with Piaget's theory where we have that the "concrete operational" stage of children cognitive development is peculiar of the primary school age. Practical experiences also facilitate in-service teachers who must be introduced to computing with short courses. As for the type of the activities to be proposed we have in mind also Martha Nussbaum's concern on the contraction of the humanistic component in the curricula contents:

> More and more often we treat education as if its primary goal should be to teach students to be economically productive rather than to think critically and to become informed and empathetic citizens [7].

The Italian National Indications for k-8 education appear to share Nussbaum's concern.

The introductory programming environment we propose for all ages is Scratch because of the reasons summarized by Shneiderman [12] and by Romeike [10] to characterize programming development environments suitable for introductory experiences.

With all the above motivations in mind, our proposal for the Scuola2.0 project was to devote the first five meetings to introduce unplugged activities plus elements of programming using Scratch and the remaining five to consider types of activities that the attendees could immediately implement in school adapted to their pupils while our lectures were continuing. Previous experiences had shown that at the end of retraining courses many of the participants had not felt able to design articulated activities adapted to their classes or, not having colleagues in the same school to share the experience with, they had been afraid to tackle alone a first field-activity. For this reason often we saw replicated some of the examples and exercises from a programming course with results not always connected to each other and with the rest of the teaching. We asked the attendees to implement one activity suitable for their students choosing out of three types of activities that we presented and discussed with them. The activity had to be developed while the second part of the meetings was going on so that the problems encountered could be discussed in person with the other attendees and the lecturer. After the end of the course, the virtual community environment of the project is still available for discussions but a starting phase where problems are discussed in persons is necessary, again from our past experiences. The three types of activities proposed are:

– inventing a narration that would gather the most relevant aspects of a topic covered in school or that might be of interest to the schoolchildren,
– inventing quizzes choosing a curriculum topic and deciding, first in group then with the entire class, a set of questions with multiple-choice answers better representing the topic,
– for middle schools: inventing riddles each modeled and solved by a linear equation.

We offered programming integrated with unplugged activities of the type suggested in [4] because we have closely worked for years with teachers in schools becoming quite respectful of their competencies in the pedagogical and methodological components of educational activities.

Here we describe how the lectures of the project went on. In Sect. 2 the general motivations are briefly resumed with a summary of the first five meetings. Sections 3 and 4 concern last five meetings. First story telling and riddles invention activities are described. Section 4 is entirely dedicated to the activities of questions & answers, organized as group and class quizzes, whose proposal and materials have been methodically organized more recently than the others.

2   Key Principles of the Project and First Meetings

In primary education activities count for what children acquire during the process of developing an activity as much as for the result produced. Digital experiences also should take into account this kind of methodological approach.

2.1   Scuola2.0 Principles

The key principles inspiring the activities suggested during Scuola2.0 are the following:

– every activity, including programming, shall be a learning environment contributing to the overall growth of the child in its ethical, social and intellectual capabilities, from the beginning to the end of the activity development,
– every action must have a specific educational goal and be integrated to the overall pedagogical and disciplinary contents of the grade it is proposed to,
– particularly in the early years, programming must be conceived as one of the "hundred" languages children shall use to create and express themselves, as from Loris Malaguzzi of Reggio Emilia schools [3].

The project could count on ten meetings, three hours each. During the first five meetings we introduced attendees to the *CS Unplugged* activities presented in [2] and to basic programming concepts, using Scratch, shortly summarized in this section.

2.2   Unplugged Programming

We have been introducing programming for years with activities that often were sort of an easier version of those present in first programming courses at the university or in technical upper secondary schools. Soon we felt mandatory to offer different activities more integrated to the educational contents and pedagogical methodologies particularly, though not only, in k-8 education. Thus we began the Teachers for teachers (T4T) experience where we work with teachers and collect suggestions from the field. We revisited Logo activities and *CS Unplugged* activities developed in schools. In [4], three primary school teachers of the T4T group describe various types of computer-related activities they have created with their pupils. For first grades of primary schools, they have experimented *CS Unplugged* activities, for example those moving a human-robot. The latter are activities on a school chessboard-like playground or similar where a pupil moves from one square to another one according to the instructions her/his mates give. Only four instructions are available at the beginning (`forward`, `backward`, `turn-left` and `turn-right`), then the instruction set is gradually enriched, for example with instructions for bringing something from a square to another one. Also pupils are requested to perform different activities such as:

– comparing the different paths obtained from different sequences of commands,
– comparing lengths of instruction sequences written by different groups.

The presence of an obstacle on the playground, in one of our schools there is a slide, enriches the possible activities since children must avoid the obstacle. Also: first writing down inside the school the instructions for a path, then verifying them on the playground, makes teachers and pupils concretely see the sequence of commands and better catch the concept. Besides, having only few lines where writing the sequence often generates the idea of parameters, `forward(n)` for example, or `repeat(n)`. This is the same ruse used in other environments, for example in Lightbot, https://lightbot.com/). Also, attendees shall find out that the human-robot written sequence of instructions corresponds to the sequence of actions we perform in some real world situations, for example similar to the sequence of actions written on the Fire Alarm Table, i.e. the actions we (must) perform when we hear the fire alarm in school. Learning achievements during unplugged programming make easier the activities that follow.

2.3   Plugged in Activities During the First Part of Scuola2.0

Like many authors recommend, programming can be present in k-8 education using an environment suitable to the age of the students. Besides, as we wrote earlier, we shall propose suitable activities. Alessandro Rabbone with his pupils developed MicroWorlds activities such as those entitled "Let's sing" ("Si canta") and "The auger" ("La trivella") during the 2004–2006 project KidsIdeasActivities (BambiniIdeeProgetti) whose final video can be seen at http://win.rabbone.it/_irreMMjr/progetti.asp#. The mentioned titles are self explaining and suggest that the relative activities are quite different from those one can find in a university or technical school course for programming. We also saw stories in Alice that some teachers had developed in secondary schools. The Scratch workshop led at ISSEP 2011 by Katarína Mikolajová and Martina Kabátová [6], and Lawrence Williams' visit to our department in 2013, who showed us several stories in Scratch [13], brought materials to our idea of changing the kind of activities we were proposing in our projects going on using Scratch.

An introduction to basic programming concepts by writing easy stories using the Scratch environment was given during our first five meetings of the Scuola2.0 project. A story telling activity allows pupils to express their creativity whether using digital tools or not. Using development environments such as Scratch this activity can be done at very different levels of familiarity with the tool, see again [13] and its references. For this reason Scratch is often proposed in courses introducing computing.

First we show on the big screen a story whose code, not considered at the beginning, is a sequence of actions only. Then we look inside the code and disassemble it asking attendees to find components of the story we just saw that is, using the theatre metaphor, they find actors/sprites, backgrounds, the costumes changed by the actors, the songs that are produced. Attendees start doing something of their own by changing costumes, dialogues. Then they continue with reviewing the synchronization among actions and so on according to the principle of remixing recommended by Resnick and the group of researchers authors of Scratch [9]. While developing stories, basic programming principles are recalled from previous *CS Unplugged* activities or are newly introduced together with some achievements from the actual use of the tool:

1. command sequences,
2. very simple repeat (repeat n times), typically to move a sprite or changing backdrops one after the other,
3. synchronization using seconds (because it is the easier to begin with) by designing a timeline of the story,
4. some interactions, for example to ask the user's name in order to personalize the execution of an activity.

The "story telling" pattern is suitable also for schoolchildren who can barely read and write and is interesting because it can smoothly evolve toward stories requiring a long time for the design and for planning the several activities to produce the narration such as the drawing of the sprites and of the backdrops, deciding the dialogues, and so on.

During the interval between the first and the second part of the Scuola2.0 meetings, some teachers were able to develop *CS Unplugged* activities with their pupils. They recognized patterns of commands used in those activities within the Scratch scripts and were more confident than the other teachers in reading the scripts of the first Scratch stories.

3   Long Activities to Be Proposed in Schools

Revising a curriculum trying to maintain a satisfying coexistence among old and new contents is a difficult task. Those who propose to introduce digital competencies in education assume a great responsibility on the one hand with respect to the contents of the other disciplines that are declined to make way for new contents, on the other hand with respect to time and money resources that are diverted to the new activities. In choosing the types of activities to be proposed during our project one of our intents was to conciliate the contents already present in k-8 education with the digital activities proposed to teachers and students. The first two activity-frameworks we proposed allow to begin with very simple activities yet introducing some programming principles and then continue with gradually increased complexity.

3.1   The "Story Telling" Pattern

As we already said, the "story telling" pattern is interesting because it can smoothly evolve from short plain narrations, sequential in their digital implementation, toward long stories involving an entire class as the "Red Riding Hood" tale produced in Scratch by fourth grade children [4]. This activity had many educational components equally important as the acquisition of computer skills. Think of the design components, the planning, the collaborative work, the definition and organization of contributions, the timing and verification of the results, the children's feeling of responsibility for finishing in time the work assigned to them: all this next to the digital implementation of the story [4]. But here we shall focus on the five meetings whose goal was to make attendees design and implement a Scratch activity with their schoolchildren while attending the project. Thus what happened in schools and possible problems could be discussed and solved with the other teachers and the lecturer.

Obviously, the discussions during the second phase of the project also gave way to enrich the knowledge and experience on programming acquired during the first meetings, in particular to solve the problems arising from the inventiveness of the schoolchildren (beginning with the typical in the field programming problems: for example those concerning how to delete a figure background to create a new sprite). The Scuola2.0 meetings had the role of organizing the different activities, defining new steps, with colleagues' suggestions, and receiving help with respect to the problems found while developing such activities in the school.

Some forms of interaction in story telling often lead to the idea of developing quiz activities. In this way we have a smooth transition from storytelling and easy types of quiz activities. Which also means a smooth introduction of variables for storing scores or remembering errors.

One of the narration activities was developed by a teacher with her second grade pupils. It concerns animals and environments where they live. Children decided to have: the house, the forest, the sea and the savanna and drew appropriate backdrops. They also drew some animals for each environment, for example a red fish for the house, a lion for the savanna. First idea was to have animals appearing on the screen, each with the proper backdrop, saying something about its life. But the teacher during our meetings liked better to have an activity to practice English. Thus she and a second teacher implemented a quiz where, when an animal appears on the screen, the child sitting in front of the screen enters the animal's English name. For this first attempt children drew the backdrops and the animals. Thus the result is only partially developed with children. But for the Scuola2.0 project we consider positive that a teacher who knew nothing of programming at the beginning of the project, after six/seven meetings introducing her to computing had the initiative of proposing to her pupils this activity. She had to organize her schoolchildren deciding with them the four environments, and then divide pupils in groups, each group working on four animals one for each setting. This teacher's idea is to ask pupils to modify current year activity letting pupils dig into the code and producing something of their own at the beginning of the next year when they will be in their third grade.

Another Scuola2.0 attendee worked with his pupils on a long story about myths of ancient Greece with different components each developed by a different group of pupils. During our meetings we discussed about an easy way of putting together the components and together we found how. Among problems to be solved we had the question of global variables, i.e. variables every sprite can see, whose copies were all maintained when the projects developed by each group were integrated in a single activity. All these conquests came from working contemporarily in schools and with colleagues during the Scuola2.0 meetings.

Going toward a quiz activity, even if very simple like the one on animal names, programming concepts introduced are:

1. *selection*: for verifying the answers,
2. *variables*: introduced if we want to count the score,
3. *repeat until condition*: possibly introduced depending on the type of quiz (in the case of the animal names, *repeat until* "the answer specifying the name of the animal is correct").

3.2   Inventing Riddles

Retraining courses using Scratch are very well received by teachers because they appreciate the use of a simple tool through which students can get rewarding results. Also teachers appreciate they are asked to work on activities fit to their students and with contents that can be interesting for other disciplines and then for fellow teachers. Examples of interdisciplinary activities are the programs "think a number (and I guess it)" in which each group of students invents its own riddle through an experimental activity on linear equations. Here we refer to a type of riddle popular in our country played between two students s1 and s2. An example of this kind of riddle is the following list of requests:

1. s1 to s2: - think a number (let's call it x)
2. s1 to s2: - add 7
3. s1 to s2: - multiply by 3
4. s1 to s2: - subtract twice the number you thought
5. s1 to s2: - add 4
6. s1 to s2: - finally divide by 5
7. s1 to s2: - what number you end up with?
8. *s2 to s1*: - I have 9
9. s1 to s2: - thus, you thought 20!

This riddle corresponds to the equation:

$$(3(x + 7) - 2x + 4)/5 = 9$$

then $3x + 21 - 2x + 4 = 45$ and $x = 20$.

We can write a Scratch activity where the student s1 is a Scratch sprite.

Let us call a the answer in 8 of student s2 to the question in 7 of student s1. In such activity $x = 5a - 25$. With answer 9 at point 8 we have $x = 5 \cdot 9 - 25$.

But you can also have another type of requests where x is cancelled. All requests are from student s1 to student s2 as in the following example:

1. think a number (again we call it x)
2. add 7
3. multiply by 2
4. subtract 4
5. subtract twice the number you thought
6. finally divide by 5
7. You got 2! How come I can guess right?

This riddle corresponds to the expression, where variable x disappears:

$$(2(x + 7) - 4 - 2x)/5$$

and $10/5$ is the final value.

In both cases the corresponding Scratch activities are very simple: each is a sequence of instructions for conducting the dialogue. The valuable part of the activity is once again, and particularly in this case, the phase where the students begin playing the riddle unplugged, are asked to build a riddle of their own which can be done when they understand that a riddle can be modeled by an equation. And then they invent other riddles.

4   Questioning

The questioning activity has been suggested by various sources in the pedagogical literature where questioning techniques are largely discussed [10]. Not least the fact that a Scratch activity designed as an exercise to introduce the variables (in this case the "score" variable) in a quiz has been very well received in all courses in which it was proposed because considered appealing for schoolchildren. Discussions with teachers, involved in the Scuola2.0 project and outside, has also shown that a type of very simple exercise such as a quiz has computing value in introducing a gradual use of the variables but also offers original ways of learning: each group of students creates its own quiz where questions are proposed on the topics of a lesson that most affected the group. These are examples of the many activities that promote the involvement of teachers in other disciplines and thus the gradual upgrade of the digital skills of these teachers also: involvement essential if you want to get to use the digital as a tool for constructively learning various disciplines.

The questioning activities we propose will have two phases:

Group work: with groups of two or three pupils per group. During this step each group produces a quiz with 3 or 4 questions on a topic, that can be a curricular one. We chose to suggest having multiple answers to increase the work by each pupil in the group. During this phase the members of the group review the topic and decide which are its representative components candidate to become matter of queries. Similarly the relevant answers to the chosen questions are decided to build a multiple-choice quiz. The group-quiz is finished with images and sounds;

Class work: at school each group shows to the entire class questions and answers present in its group-quiz. The goal is to produce a class-quiz from group-quizzes with more questions. Questions and answers are chosen among those proposed in the group-quizzes, both can result from different formulation of what contained in group-quizzes to take into account aspects deemed important by the children or according to teachers' suggestions.

The class-quiz can have various uses: for example, can be shared with another class having the same subject in curriculum. This will assess the pupils' ability both to answer questions and to evaluate possibly missing important aspects of the considered topic.

Positive aspects of this activity are:

- the discussions that are developed on the topic,
- the active learning aspect,
- the possibility to involve all students. In the digital implementation maybe you can choose to put in the class-quiz at least the figures of a group or of a student disappointed with respect to questions/answers, for example because the ones he has proposed have been discarded.

Working to build a quiz stimulates a series of reflections and activities that involve different learning areas. As in the story telling, the planning phase requires the choice of a topic and the definition of the objectives, the selection of materials and the definition of tasks and deadlines, then the organization of a scrum board. Later and during the development of the activity, critical comparisons are necessary and then, a self evaluation test at the end of the activity on the strengths and critical points of the process and of the result. Figure 1 shows a screenshot of a quiz on neurons (from an activity by Carlotta Craveri, student of Educational Sciences in our university during the academic year 2014–15). The figure shows the question "How do neurons communicate?" whose proposed answers are:

A. Using smoke signals
B. Using WhatsApp
C. By means of the synapses

This example has been shown to the attendees of the Scuola2.0 meetings. In fact for each activity type, because of the short time available, it was decided to show and discuss with the teachers a framework, i.e. a very simple yet running example for each one of the proposed activity, and then let them start working on such frameworks.

Fig. 1. How do neurons communicate?

5   Conclusions

This is not a contribution toward defining digital competencies needed to future k-8 teachers. It is a contribution to deciding how in-service teachers can be trained to gradually gain the competencies for developing with their pupils the digital activities we need to offer in k-8 education without further delay. Of course it requires we define, at least partially, what we consider the proper digital curriculum for k-8 education so that we can go toward the chosen direction.

Here we shortly report what has been the computing training for a group of k-8 teachers. The paper particularly concerns the second half of the training meetings where we pushed the teachers to develop with their schoolchildren an activity lasting for the two last months of the school year (April and May in our country because June is to be left for conclusive verification activities). The teachers were suggested three types of activities aiming to have all attendees working on a program that lasted for several weeks.

The seventeen primary school teachers attending the Scuola2.0 meetings have all chosen the story telling activity that is in fact an activity feasible at very different levels of difficulty: one can start by modifying a given story that, in our case, is the given reference framework. Instead, teachers in middle schools judged feasible in their classes all three kinds of activities but, perhaps because they are all mathematicians, they focused on stories and riddles (the latter also because more directly related to mathematics) motivating the choices with the short time available in this end of school year 15/16.

The operating methodology we chose provided the course attendees with a framework of each of the suggested types of activity, i.e. a working simplified version of it. According to the participants this method is proving to be very useful. Also useful the Moodle community where participants could ask for help with problems and discuss solutions during the two months when the meetings took place. The community is almost essential after the end of the course. Not all teachers could attend the last meetings and only few of the attendees could actually develop almost entirely the chosen activity with their pupils. From the meetings assessment, the teachers motivated this failure with their short and busy time when the end of the school year is approaching. As we wrote, digital activities in primary school are learning environments to be entirely developed from the design phase without aiming for the final result only. Thus the request of the teachers to have all the meetings in the very beginning of the school year could be a way to increase the actual transfer of their experience inside their schools.

We do not have an assessment of our experience yet, but we have seen teachers curious of implementing something that on the one hand they felt near and useful for what they were doing every day, on the other hand new and fitting the direction schools shall take in the future. Also, the teachers considered quite positively the narrative and the question/answer frameworks as a continuation of each other. This was perceived as having an activity to develop with pupils lasting for several sessions and consequently not an occasional exercise that is one of their frequent complaints. The second half meetings turned out to largely be a discussion time for solving implementation problems, for reciprocally showing the activities, exchanging ideas and pedagogical comments, acquiring new abilities from colleagues' suggestions. The general feeling was that the activities developed during the meetings helped overcoming the teachers' fear of inventing something of their own. The integration of educational content specific to the class where an activity is proposed is an outcome of the active involvement of the teachers.

An open question concerns the MOOC with the same contents of the course here described, that the Municipality recently requested us. Our doubts on its effectiveness come from considering that the in person meetings gave a fundamental contribution to the success of this course.

References

1. Ben-Ari, M.M.: In defense of programming. ACM Inroads **7**(1), 44–46 (2016). doi:10.1145/2827858
2. Berry, M.: Computing in the national curriculum. A guide for primary teachers. Computing at School (2013). http://www.computingatschool.org.uk/data/uploads/CASPrimaryComputing.pdf
3. Cagliari, P., Castagnetti, M., et al.: Loris Malaguzzi and the Schools of Reggio Emilia. Routledge, Abingdon (2016)
4. Ferrari, F., Rabbone, A., Ruggiero, S.: Experiences of the T4T group in primary schools. In: Jekovec, M. (ed.) The Proceedings of International Conference on Informatics in Schools: Situation, Evolution and Perspectives - ISSEP 2015. University of Ljubljana, Faculty of Computer and Information Science, October 2015
5. Gander, W., et al.: Informatics education: Europe cannot afford to miss the boat. Report of the Joint Informatics Europe & ACM Europe Working Group on Informatics Education, ACM Europe, April 2013. http://europe.acm.org/iereport/ACMandIEreport.pdf
6. Kabátová, M., Mikolajová, K.: Fostering creativity through programming - scratch workshop. In: Bezáková, D., Kalaš, I. (eds.) ISSEP 2011 Proceedings of Selected Papers. Comenius University, Faculty of Mathematics, Physics and Informatics, Bratislava, October 2011. http://www.issep.2011.org/files/workshops/w07.pdf
7. Nussbaum, M.C.: Not for Profit: Why Democracy Needs the Humanities. Princeton University Press, Princeton (2010)
8. OECD: Students, Computers and Learning. OECD Publishing (2015). http://dx.doi.org/10.1787/9789264239555-en
9. Resnick, M., Maloney, J., Monroy-Hernández, A., Rusk, N., Eastmond, E., Brennan, K., Millner, A., Rosenbaum, E., Silver, J., Silverman, B., Kafai, Y.: Scratch: programming for all. Commun. ACM **52**(11), 60–67 (2009). doi:10.1145/1592761.1592779
10. Romeike, R.: Three drivers for creativity in computer science education. In: Benzie, D., Iding, M. (eds.) Proceedings of the Working Joint IFIP Conference Informatics, Mathematics, and ICT: a 'golden triangle', IMCT 2007, Boston (2007)
11. Schulte, C.: Reflections on the role of programming in primary and secondary computing education. In: Proceedings of the 8th Workshop in Primary and Secondary Computing Education, WiPSE 2013, pp. 17–24. ACM, New York (2013). http://doi.acm.org/10.1145/2532748.2532754
12. Shneiderman, B.: Creativity support tools: accelerating discovery and innovation. Commun. ACM **50**, 20–32 (2007)
13. Williams, L., Černochová, M.: Literacy from scratch. In: Proceedings of the 10th IFIP World Conference on Computers in Education, WCCE 2013, pp. 17–27. Copernicus University, Torun (2013)

Defining and Observing Modeling and Simulation in Informatics

Nataša Grgurina[1(✉)], Erik Barendsen[2], Bert Zwaneveld[3],
Klaas van Veen[1], and Cor Suhre[1]

[1] Teaching and Teacher Education, University of Groningen,
Groningen, The Netherlands
{n.grgurina,klaas.van.veen,c.j.m.suhre}@rug.nl
[2] Radboud University and Open University, Nijmegen, The Netherlands
e.barendsen@cs.ru.nl
[3] Open University, Heerlen, The Netherlands
g.zwaneveld@uu.nl

Abstract. Computational Thinking (CT) is gaining a lot of attention in education. In this study we focus on the CT aspect modeling and simulation. We conducted a case study analyzing the projects of 12th grade high school informatics students in which they made models and ran simulations of phenomena from other disciplines. We constructed an analytic framework based on literature about modeling and analyzed students' project documentation, recordings of student groups at work and during presentations, survey results and interviews with individual students. We examined how to discern the elements of our framework in the students' work. Moreover, we determined which data sources are suitable for observing students' learning. Finally, we investigated what difficulties students encounter while working on modeling and simulation projects. Our findings result in an operational definition of modeling and simulation, and provide input for future development of both assessment instruments and instructional strategies.

Keywords: Computational thinking · Modeling and simulation · Informatics · Secondary education

1 Introduction

Following the increasing availability of computers in schools, several initiatives have been employed to aid students' learning in various disciplines through the use of computer models [2, 23, 24]. Nowack and Caspersen [6] argue why they *"believe understanding and creating models are fundamental skills for all pupils as it can be characterized as the skill that enable us to analyze and understand phenomena as well as design and construct artifacts."* Wilensky argues, *"Computational modeling has the potential to give students means of expressing and testing explanations of phenomena both in the natural and social worlds"* [27]. Granger claims: *"Modeling is the new literacy"* [9]. This belief is also expressed in the fact that as of 2019, modeling and simulation (together called *Computational Science*), will be included in the new Dutch

© Springer International Publishing AG 2016
A. Brodnik and F. Tort (Eds.): ISSEP 2016, LNCS 9973, pp. 130–141, 2016.
DOI: 10.1007/978-3-319-46747-4_11

high school informatics curriculum, described by the following high level learning objectives: *"Modeling: The candidate is able to model aspects of a different scientific discipline in computational terms"* and *"Simulation: The candidate is able to construct models and simulations, and use these for the research of phenomena in that other science field."* Modeling itself will be a part of the compulsory core curriculum, described as *"Modeling: The candidate is able to use context to analyze a relevant problem, limit this to a manageable problem, translate this into a model, generate and interpret model results, and test and assess the model. The candidate is able to use consistent reasoning"* [1].

Modeling and simulation can be viewed as aspects of *Computational Thinking* (CT) [30] as they involve decomposition of open-ended problems and the construction and evaluation of models that simulate the nature of these problems in order to be able to provide solutions to those problems. The present study is part of a larger research project on CT in Dutch secondary education, investigating pedagogical aspects of CT and teachers' pedagogical content knowledge (PCK, [22]) about these aspects. Following Magnusson et al. [17], we distinguish four aspects of content-specific pedagogy: (1) goals and objectives, (2) students' understanding and difficulties, (3) instructional strategies, and (4) assessment. In the first phase of the project, we refined the CSTA definition of CT [10], explored teachers' PCK [11, 12]; and made an initial exploration of the computational modeling process [13].

Aim of the Study. In this study we focus on CT skills related to modeling and simulation and we explore highly cognitively complex set of students' activities related to modeling, in particular as an aspect of CT rather than as an aspect of e.g. mathematics [16]. We address the following research questions:

1. How can the intended learning outcomes of *Computational science* (modeling and simulation) be described in operational terms?
2. What data sources are suitable to monitor students' learning outcomes when engaging in modeling activities?
3. What specific challenges do the students experience when engaging in modeling activities?

The first question addresses Magnusson's aspect (1). The second contributes to aspect (4) – we plan to use our findings as input for a later study into a CT assessment instrument. The third question addresses aspect (2); our findings will help design teaching materials for modeling and simulation and thus indirectly contribute to (3).

Related Work. Previous work on characterizing modeling is done mainly in the areas of mathematics and natural sciences; see the following section. Research on making students' learning process and outcomes visible has focused mostly on CT aspects such as algorithmic thinking or programming. The employed assessment instruments range from tests with closed questions [8], tests with open questions [18, 26], surveys [26], recordings or observations of students at work [18], examination of programming projects [4, 18, 26] to interviews with students [4, 14] and teachers [18]. In particular, Brennan and Resnick [4] *"are interested in the ways that design-based learning activities [...] support the development of computational thinking in young people"* and explore three approaches to assessment of the development of CT of the children engaged in such activities. They discuss strengths and limitations of each of these approaches extensively and subsequently advocate a comprehensive approach to assessment that utilizes several data sources.

Context of the Study. Our exploratory case study was carried out during a project-based lesson series within a regular informatics course in the 12th grade of high school where students studied modeling and simulations. They used NetLogo to program models of phenomena from other disciplines and to explore them through running simulations. During a six-weeks period they studied Modeling and Simulations with NetLogo. The first three weeks were dedicated to studying the textbook material. During the rest of the period, the fourteen students comprising this class worked in seven groups on a practical assignment where they investigated a phenomenon of their choosing by making a model in NetLogo and exploring it through running simulations. When necessary, students were assisted in formulating their hypotheses or research questions. The entire process was strictly planned and contained milestones when the students turned in the required project documentation. At the end of the period, each group presented its model to the rest of the class and the students were encouraged to discuss their models, results, design choices, programming issues and other relevant questions. After the presentations, they turned in the final part of the project documentation where they described the feedback they got and their reaction to it. A few days later, twelve students (six groups) who finished their projects, turned in their final reports and NetLogo programs.

2 Modeling and Simulation

There is extensive literature on modeling in science and especially in mathematics. We take the latter as starting point and discuss *simulation modeling* as a special case of modeling.

2.1 Modeling

Van Overveld et al. [25] distinguishes two purposes of modeling: scientific research and technological design, and lists a number of goals that can be obtained through modeling: explanation, prediction, compression, abstraction, unification, analysis, verification, communication, exploration, decision, optimization, specification, steering and control and finally: training. The mathematical modeling process can be viewed as a problem solving activity (cf. [20]). We adopt the operationalization by Van Overveld et al. [25]:

1. *Definition stage*: the problem is stated and **researched** in the context domain (this is also considered a core aspect of CT [30]). The **purpose** of the model is formulated and a study is planned.
2. *Conceptualization stage*: Data are collected and a conceptual model is constructed and validated. In the process of **abstraction** it is decided what details to highlight and what details to ignore.

3. *Formalization stage*: the conceptual model is transformed into a formal model.
4. *Execution stage*: the model is being used for its purpose: this means solving the mathematical problem.
5. *Conclusion stage*: the results of the execution stage are **analyzed** and translated back into the problem domain, involving the presentation and interpretation of the results.

In addition, Perrenet and Zwaneveld [19] explicitly distinguish between the non-mathematical world containing the definition stage, conceptualization stage and conclusion stage on the one hand, and the mathematical world containing the formalization and execution stages on the other hand. Following each of these stages, reflection needs to take place: to check if any revisions are necessary by repeating that stage, to validate and verify the model, to assess the plausibility of the result and answer the initial purpose, to communicate the results and to learn from what one has done. After the completion of the modeling process, a **reflection** takes place and the whole process is possibly repeated. Hence, modeling can be seen as a *cyclic* process [19, 25].

2.2 Simulation Modeling

Simulation modeling can be seen as a special case of modeling in which the model consists of a computer program and therefore is *executable*. In comparison to the mathematical modeling process, the simulation modeling process shows a computational – rather than mathematical – interpretation of the conceptualization, formalization and execution stages:

1. *Conceptualization stage*: Data are collected and a conceptual model is constructed and validated. In the process of abstraction it is decided what details to highlight and what details to ignore. Problem is **formulated** in a way that enables us to use a computer and other tools to help solve them [5].
2. *Formalization stage*: a computer program is constructed, i.e. **requirements** and **specifications** are stated and the system is **implemented** and tested [7]. This includes making pilot runs, **verifying** the program and checking **validity** of the simulation model. If necessary the program is adjusted [15]. Thus, the formalization stage is a cyclic process in itself.
3. *Execution stage*: the model is being used for its purpose: designing and running **experiments** [15].

Simulation modeling encompasses three methods: (1) *System dynamics*, associated with high level of abstraction where the individual objects are aggregated. The models can be described in terms of differential equations that are often non-trivial to solve. (2) *Discrete event modeling*, where the system modeled is considered to be a process, *"i.e. a sequence of operations being performed across entities"*. The level of abstraction is lower. (3) *Agent based modeling* (ABM), which is made possible with recent growth of availability of CPU power and memory, does not assume any particular abstraction level. Agents have their properties and behavior and one can start building a model by identifying agents and describing their behavior even without knowing how a system behaves as a whole. ABM makes it possible to model systems that are difficult to capture with older modeling approaches [3]. In our view, the last two characteristics of the ABM make it a suitable modeling method for our students who often lack deep understanding of the phenomena they model and make models specifically to deepen their understanding. To conclude, we consider conceptual representation which could be realized through the employment of ABM methods and software, in which *"you give computational rules to individual agents and then observe, explore analyze the resultant aggregate patterns"* [27] suitable for use in secondary informatics class *"because the individual-level behavior of agents is relatively simple, [and] ABMs feature relatively simple computer programs that control the behaviors of their computational agents"* [28].

In simulation modeling, repeating the conceptualization stage or going back and forth between the conceptualization stage and formalization stage are considered to be an integral part of the modeling process [15]. In the specific case of ABM, the boundaries between all modeling stages are blurred and it is considered a good modeling practice to develop a model in minute increments, cycling continuously through all modeling stages [29].

3   Method

The data were collected by the first author during the project based lesson series. In view of existing studies involving algorithmic thinking and programming (see the introduction), we decided to use a combination of several data sources as a promising approach for our exploration of the students' activities and learning difficulties in their projects.

During their work in the class and the final presentations, screen and voice *recordings* were made of students' groups. (No recordings were made of students working elsewhere, such as at home). Except for a few corrupted recordings, they were all transcribed verbatim. The *project documentation* of each group was collected. After receiving their grades, twelve students filled in an online *survey* individually where they were asked about how they approached the work on this project, difficulties they encountered, what they have learned, what they liked or disliked, and what suggestions they had for the improvement of the assignment. Students were also invited to be *interviewed*. Five semi-structured interviews were conducted with individual students. The students were requested to describe their projects and they were asked if they could design a new NetLogo model on the fly (i.e., draw a sketch of the interface on paper and describe the model in terms of agents and interactions). Finally, they were asked what they learned during their work on the projects. The interviews were recorded and transcribed verbatim.

Using atlas.ti CAQDAS software we performed a qualitative analysis of the recordings, project documentation, survey results, and interviews, with coding categories based on the elements of our operational definition (i.e., boldface items in Sect. 2): purpose, research, abstraction, formulation, requirements, specification, implementation, verification, validation, experiment, analysis, and finally, reflection. After coding, we ascertained the visibility of the modeling elements in each of the data sources and examined the students' activities more in-depth, looking specifically for indications of students' difficulties connected to each of the elements.

4   Results

There were seven project teams. Five teams consisted of two students; one of three students, and one student opted to work by himself. Six of seven projects were successful; Team 5 did not finish theirs and did not turn in all the required project documentation. We first present an overview of visible occurrences of the elements of our modeling operationalization, organized by data source and student (team): see Table 1. Some elements were combined – see the descriptions below.

We now summarize the findings of our more in-depth analysis, organized by the elements of our operational description. We state our findings in general terms and illustrate them with characteristic text segments taken from the data.

Purpose. In the project documentation all teams clearly stated the purpose of their models. However, in the recordings we saw students tinkering with NetLogo and looking at existing models before deciding what phenomenon they wanted to model and explore. In answering the survey question whether it was difficult to decide what phenomenon to model and explore, four students answered affirmative and told us they had difficulties figuring out what could or could not be modeled. For S4a, who explored the behavior of partygoers together with S4b, the most important lesson learned during his work on this project was that it was important to have a clear idea of the purpose of the model before engaging in the modeling process – a thought shared by three other students in the survey.

Research. In the recordings we saw three students from two groups searching the Internet to learn about the phenomena they modeled. Team 3 reported in the documentation of their project about the possibilities to control the spread of the Ebola virus: *"Virus: does not spread through the air but through contact with an Ebola patient (sex, blood), slaughtering and eating of a sick animal, non-sterile needles. […] incubation about 21 days, 9 out of 10 people die",* without reporting the source. In the survey, S3b mentioned consulting her sister who studied medicine. Team 6, exploring the effect of ambient warmth and the presence of food on life of bacteria, did not report any research in their project documentation. Others did not visibly engage in research but developed their models based on what they already knew about the phenomena they modeled (e.g. Team 1 who explored chemical reactions – in the survey S1a wrote they learned that in chemistry lessons) or their presumptions (e.g. Team 7, who explored whether mousetraps were more effective than cats in catching mice, or Team 5, who explored the influence of weather on ice cream sales).

Abstraction. All students engaged in abstracting: choosing a level of abstraction, based on the decision they made with respect to relevancy of particular features and deciding what to include into their models and what to leave out.

In the recording we observed several students struggling to determine such a level of abstraction. For example, Team 1 - who initially neglected teacher's instruction to

Table 1. Frequencies of simulation modeling elements per data source per team or student. For example, Team 3 consists of students S3a and S3b.

		Purpose	Research	Abstraction	Formulating	Requirements and specification	Implementation	Verification and Validation	Experiment	Analysis	Reflection
Project documentation	Team 1	✓	✓	✓	✓	✓		✓			✓
	Team 2	✓		✓	✓	✓		✓			✓
	Team 3	✓	✓	✓	✓	✓	✓	✓		✓	✓
	Team 4	✓		✓	✓	✓		✓	✓		✓
	Team 5	✓		✓							
	Team 6	✓		✓	✓	✓		✓	✓		✓
	Team 7	✓	✓	✓	✓	✓	✓	✓	✓	✓	✓
Surveys	S1a			✓		✓		✓	✓		✓
	S1b								✓		✓
	S1c	✓	✓			✓	✓		✓		
	S2	✓	✓	✓	✓		✓	✓	✓		✓
	S3a	✓	✓	✓	✓		✓			✓	✓
	S3b	✓	✓	✓			✓	✓	✓		✓
	S4a	✓						✓			✓
	S4b	✓	✓	✓	✓		✓	✓	✓		✓
	S5a	✓	✓				✓		✓		✓
	S6a		✓								
	S7a		✓	✓		✓	✓		✓		✓
	S7b		✓	✓					✓		
Interviews	S1a	✓		✓	✓	✓	✓	✓	✓		✓
	S2	✓	✓	✓	✓	✓	✓	✓	✓	✓	✓
	S3a	✓	✓	✓		✓	✓			✓	✓
	S3b	✓	✓	✓	✓	✓	✓	✓	✓	✓	✓
	S7a	✓		✓	✓	✓	✓	✓	✓	✓	✓
Recordings	Team 1	✓	✓		✓	✓	✓		✓		
	Team 2	✓		✓	✓		✓				
	Team 3	✓	✓	✓	✓	✓	✓	✓	✓	✓	✓
	Team 4	✓	✓	✓		✓	✓	✓	✓	✓	✓
	Team 5	✓		✓	✓	✓	✓	✓	✓	✓	✓
	Team 6	✓	✓	✓		✓		✓	✓	✓	
	Team 7	✓	✓	✓	✓		✓				
	Presentations	✓		✓		✓	✓	✓	✓	✓	✓

study the textbook first - had difficulties understanding the idea behind ABM and got 'stuck' in the notion of an aggregate state, e.g. thinking about pH as a contributing factor in a chemical reaction rather than the result of it. During the interview, S7a told us that he wanted his mice to reproduce but did not include this feature because he did not know how to implement males and females. It did not occur to him that gender of the mice was not relevant in his model. Finally, as required, all students who finished the project turned in wish lists with features or aspects that were not implemented yet but should be considered for the next version of the model, thus demonstrating they were able to decide what to include or leave out.

Formulation. The assignment required a description of the behavior of the model in a natural language, and all the students who finished their projects did that. However, several students needed help to formulate their problems appropriately: e.g. only after choosing the right level of abstraction did Team 1 manage to formulate their problem appropriately and in the recordings we heard S1a say, *"Two of these things have to collide with each other and then something needs to happen".*

Requirements and Specification. In the recordings it turned out to be hard to observe a distinction between requirements and specifications – see the results on Testing for a comprehensive example.

A description of requirements and specifications was a part of project documentation and all the students who finished their projects provided it. Team 7 wrote, *"The mousetraps need to be placed at random locations since we don't know what the perfect locations would be. If a mouse contacts a mousetrap, then the mouse needs to die/disappear."* In their project documentation, Team 1 stated requirements: *"In our program, two particles react to form two other particles. The probability that two particles react can be specified, as well as the reaction speed of the particles."* Then they wrote specifications extensively: *"If two initial turtles (red and yellow) meet, then the current catalyst value (the left slider) determines whether they react."* Team 3 wrote, *"In our model there is only one [breed of] turtles and it stands for people. These turtles can have various properties, such as being ill or healthy. They can be influenced by external factors such as medicine and their life span."*

Implementation. While all students managed to implement something, some of them experienced difficulties. In the recordings we heard S1a say, *"I know what I want to do, but I don't know how to code it. I don't think it's all that difficult, but..."* During the interview, S7a told us he refrained from including mouse reproduction because he did not know how to design this feature and program it. When constructing their programs, only one team worked top-down: the others rather engaged in bottom-up incremental development constantly adding new features to their models.

Verification and Validation. The recordings revealed a complex picture in which the distinction between validation and verification was not always clear. Team 4's approach is representative of students' strategy: they constructed their model (program) by cycling among stating requirements and specifications, implementing and testing, in minute steps: *"We have to do that with time, man, that they can only drink one beer in ten seconds or so, otherwise they drink too much!"* When testing, it was not clear whether they were validating their model or verifying their code: often they would run their program, see remarkable behavior and subsequently change the code. S4b: *"All dead."* S4a: *"It begins to deteriorate now [in the simulation, the beer is gone and people leave* *the party quickly]"* S4b: *"But how could they all get the same amount of beer?"* S4a: *"That's because of that piece of code."* S4b: *"Really? Can't that be changed? How did they do it with the sheep? [Referring to an example from NetLogo's models library]"* Subsequently they would change their code and continue their work in a similar fashion. Team 6 worked similarly. It was not clear whether S6b was validating or experimenting: *"It works now but it is not balanced, so to speak."* S6a: *"Yeah."* S6b concluded: *"Yeah. That remains to be done"* and went on to change the code. Later on they tried again. S6b: *"And if we make this one a bit lower, say seven or so, then they die, that is really abrupt, like, either they live or so, or all dead."*

In the project documentation, all of the students reported that their models behaved as expected (validation). Several students described validating their models and adjusting when necessary. To this end, Team 1 wrote: *"to prevent particles from reacting with each other immediately following a reaction, we built in a reaction pause. [...] That way you prevent particles from being stuck in a constant forth-and-back reaction."*

Experiment. Team 7 was the only team who documented systematically performed experiments with their model: they reported the initial parameter values (e.g. ten cats and nine mice) and included the resulting data plots in their project documentation. In the recordings we saw other students engage in experimenting to various degrees, but most failed to mention this in the project documentation.

Analysis. Not all the students provided an analysis of the results of the experiments, but in the project documentation, they all reported answers to the purpose of their models. Team 7's analysis revealed, *"The mousetraps were not always effective. Some mousetraps go off but the mouse manages to escape."* Finally, they concluded that mousetraps were more effective than cats in catching mice. In the recordings we saw Team 3 analyze their data, without reporting it in the project documentation, and their conclusion was, *"We expected that the new medicine would decrease the spreading of Ebola. It turned out that the medicine worked rather quickly, but that the rate of infectiousness was of influence as well."*

Reflection. As required, all the students reflected on their models in the project documentation. Team 7 wrote: *"Not everything in our model corresponds with the reality. But it is nice to experiment with it. You can make your model as large and complex as you wish."* In the survey the students were asked what they learned. S3b wrote: *"It [modeling] is a good means to predict/research hypotheses. A good aid for research. I take chemical reactions as an example. You can make it and thus see (visualize) what happens,"* a thought reflected by S1a too. S4a learned that it was important to have a clear idea of the purpose of the model before engaging in the modeling process and that models and simulations never completely correspond to the reality. Contrary to S4a's reply, during the interview S1a expressed his astonishment about how easy it was to make a model that *"actually reasonably corresponded"* to what was modeled. He even went on to show it to his chemistry teacher.

5   Conclusion and Discussion

As to the **first research question** – *How can the intended learning outcomes of Computational science (modeling and simulation) be described in operational terms* – we have obtained an operational description based on literature on modeling and simulation. The elements of the description turned out to be suitable to classify simulation modeling activities of the students in our study. However, some of these had to be grouped together since the separate elements could not be distinguished. The resulting operationalization contains the elements *purpose, research, abstraction, formulation, requirements/specification, implementation, verification/validation, experiment, analysis, reflection*. This 'blurring' of activities is also described by Wilensky and Rand [29].

In answering our **second research question** – *What data sources are suitable to monitor students' learning outcomes when engaging in modeling activities* – we found that every source enabled us to observe some aspects of the modeling process. The interviews provided the opportunity to observe all the aspect of the modeling process, closely followed by recordings of students at work. In the project documentation, the description of the model and the reflection are well represented and experimenting and analysis not so: contrary to the presentations, where exactly the opposite happens. The surveys, in their present form, did not provide much insight into the modeling stages the students engage in.

We are planning to use our results to develop an assessment instrument. In order for such an instrument to be feasible for classroom usage, a combination of project documentation and class presentation are promising data sources that enabled us to capture all aspects of students' work. Our findings suggest that the instructions for documentation and presentation could be sharpened to improve visibility of (systematic) experimentation and data analysis within the model.

Finally, in answering our **third research question** – *What specific challenges do the students experience when engaging in modeling activities* – we identified some difficulties. Many students could not decide what to model exactly, and found it hard to decide on the level of abstraction and formulate the problem suitably for modeling through ABM. While all students managed to program something, not all of them were able to program all they wanted because either they could not decide on the relevance of a feature, or they did not know enough NetLogo to code it. During testing it appeared to be difficult to attribute unexpected behavior to a fundamental modeling mistake, a programming error, or unexpected (i.e. *emergent*) behavior that was characteristic for the phenomenon under scrutiny. Students tend to rely on an incremental trial-and-error strategy while implementing their simulation model. Only a few conducted systematic and well documented experiments. Most of these experiments, together with the analysis of the results, were intermingled with the construction of the models.

This incremental development is consistent with description of the modeling practice, for example by Wilensky and Rand [29]. An ad hoc incremental development (trial-and-error strategy) is typical for novices [21].

General Remarks. Although this was a small study with a limited number of participants, we learned a lot about students' understanding of modeling and simulation.

Also, our findings indirectly informed us about the quality of the instruction, which leaves room for improvement.

Several students told us that through work on this project, they learned about the phenomena they modeled, which is in line with earlier findings [2, 24]. We often heard them laugh during their work and we observed that many students enjoyed working on this project. We saw that these informatics students were able to utilize their informatics/CT knowledge and skills to advance their learning in other disciplines.

In conclusion, we believe that the results of this research will contribute to the development of (1) suitable learning activities both within the informatics courses as elsewhere and (2) knowledge about teaching, monitoring and assessment of the CT aspect modeling and simulation. They will contribute to the development of the informatics curriculum in secondary education in the Netherlands, informatics teacher training and informatics education in general.

Acknowledgments. This work is supported by The Netherlands Organisation for Scientific Research grant nr. 023.002.138.

References

1. Barendsen, E., Tolboom, J.: Advisory Report (Intended) Curriculum for Informatics for Upper Secondary Education. SLO, Enschede (2016)
2. Blikstein, P., Wilensky, U.: An atom is known by the company it keeps: content, representation and pedagogy within the epistemic revolution of the complexity sciences (2009)
3. Borshchev, A.: The Big Book of Simulation Modeling: Multimethod Modeling with AnyLogic 6. AnyLogic North America, Chicago (2013)
4. Brennan, K., Resnick, M.: New frameworks for studying and assessing the development of computational thinking (2012)
5. CSTA Computational Thinking Task Force. Operational Definition of Computational Thinking for K-12 Education. http://csta.acm.org/Curriculum/sub/CurrFiles/CompThinkingFlyer.pdf. Accessed 16 Oct 2013
6. Caspersen, M.E., Nowack, P.: Model-Based Thinking & Practice
7. Comer, D.E., Gries, D., Mulder, M.C., Allen Tucker, A., Turner, J., Young, P.R., Denning, P.J.: Computing as a discipline. Commun. ACM **32**(1), 9–23 (1989)
8. Gouws, L., Bradshaw, K., Wentworth, P.: First year student performance in a test for computational thinking. ACM, East London, South Africa (2013)
9. Granger, C.: Coding is not the new literacy. http://www.chris-granger.com/2015/01/26/coding-is-not-the-new-literacy/. Accessed 09 Oct 2015
10. Grgurina, N.: Computational thinking in Dutch secondary education (2013)
11. Grgurina, N., Barendsen, E., Zwaneveld, B., van Veen, K., Stoker, I.: Computational thinking skills in Dutch secondary education: exploring pedagogical content knowledge. ACM (2014)
12. Grgurina, N., Barendsen, E., Zwaneveld, B., van Veen, K., Stoker, I.: Computational thinking skills in Dutch secondary education: exploring teacher's perspective. ACM (2014)
13. Grgurina, N., Barendsen, E., van Veen, K., Suhre, C., Zwaneveld, B.: Exploring students' computational thinking skills in modeling and simulation projects: a pilot study. ACM (2015)

14. Grover, S.: Robotics and engineering for middle and high school students to develop computational thinking (2011)
15. Law, A.M.: Simulation Modeling and Analysis, 5th edn. McGraw-Hill, New York (2015)
16. Maaß, K.: What are modelling competencies? ZDM Math. Educ. **38**(2), 113–142 (2006)
17. Magnusson, S., Krajcik, J., Borko, H.: Nature, sources, and development of pedagogical content knowledge for science teaching. In: Gess-Newsome, J., Lederman, N.G. (eds.) Examining Pedagogical Content Knowledge, pp. 95–132. Springer, Netherlands (1999)
18. Meerbaum-Salant, O., Armoni, M., Ben-Ari, M.: Learning computer science concepts with scratch. Comput. Sci. Educ. **23**(3), 239–264 (2013)
19. Perrenet, J., Zwaneveld, B.: The many faces of the mathematical modeling cycle. J. Math. Model. Appl. **1**(6), 3–21 (2012)
20. Polya, G.: How to Solve It: A New Aspect of Mathematical Method. Princeton University Press, Princeton (2008)
21. Robins, A., Rountree, J., Rountree, N.: Learning and teaching programming: a review and discussion. Comput. Sci. Educ. **13**(2), 137–172 (2003)
22. Shulman, L.S.: Those who understand: knowledge growth in teaching. Educ. Res. **15**(2), 4–14 (1986)
23. Spodniakova Pfefferova, M.: Computer simulations and their influence on students' understanding of oscillatory motion. Inform. Educ. **14**(2), 279–289 (2015)
24. Taub, R., Armoni, M., Ben-Ari, M.M.: Abstraction as a bridging concept between computer science and physics. ACM (2014)
25. Van Overveld, K., Borghuis, T., van Berkum, E.: From problems to numbers and back. In: Lecture Notes to 'A Discipline-Neutral Introduction to Mathematical Modelling'. Eindhoven University of Technology, Eindhoven (2015)
26. Werner, L., Denner, J., Campe, S., Kawamoto, D.C.: The fairy performance assessment: measuring computational thinking in middle school. ACM, Raleigh, North Carolina, USA (2012)
27. Wilensky, U.: Computational thinking through modeling and simulation. White paper Presented at the Summit on Future Directions in Computer Education, Orlando, FL (2014). http://www.stanford.edu/~coopers/2013Summit/WilenskyUriNorthwesternREV.pdf
28. Wilensky, U., Brady, C.E., Horn, M.S.: Fostering computational literacy in science classrooms. Commun. ACM **57**(8), 24–28 (2014)
29. Wilensky, U., Rand, W.: An Introduction to Agent-Based Modeling: Modeling Natural, Social, and Engineered Complex Systems with NetLogo. MIT Press, Cambridge (2015)
30. Wing, J.M.: Computational thinking. Commun. ACM **49**(3), 33–35 (2006)

K-12 Computer Science Education
Across the U.S.

Hai Hong, Jennifer Wang$^{(\boxtimes)}$, and Sepehr Hejazi Moghadam

1600 Amphitheatre Parkway, Mountain View, CA 94043, USA
jennifertwang@google.com

Abstract. Our multi-year national research study examines knowledge and perceptions of computer science (CS), disparities in access, and barriers to offering CS in the United States. The first year of the study surveyed 1,673 students, 1,685 parents, 1,013 teachers, 9,693 principals, and 1,865 superintendents, and the second year surveyed 1,672 students, 1,677 parents, 1,008 teachers, 9,244 principals, and 2,227 superintendents. We found that while large majorities of respondents from all groups continue to hold positive perceptions of computer science work as fun, exciting, and socially impactful, perceptions of who can do CS remained narrow. Despite support from large majorities in all groups for having CS in schools, few teachers or administrators strongly agree that CS is a top priority in their school or district, and principals report mixed support for CS from key stakeholders. Few principals and superintendents describe demand for CS from students and parents as high, while few parents and teachers report having specifically expressed support for CS education to school officials. Our paper also uncovers overall opportunities to learn CS in- and out-of-school. We see an increase in the percent of schools teaching computer programming/coding. Even if opportunities exist, students and parents may not know about them; just over half of students and teachers and 43 % of parents are aware of CS learning opportunities in the community, with slightly higher percentages of students and parents aware of online opportunities. Barriers to offering CS in schools remain largely unchanged from year one of the study, with principals continuing to cite a lack of teachers with the necessary skills and a prioritization of courses related to testing requirements as reasons why CS is not offered in their schools. To overcome such barriers, we discuss a potential opportunity for teachers to incorporate CS into existing school subjects.

Keywords: K-12 · Pre-university · Girls · Gender · Underrepresented · Black · Hispanic · School · Student · Parent · Teacher · Principal · Superintendent · Technology · Computational thinking

1 Introduction

Technical advancements and the expansion of professions in which computer science (CS) is relevant make it more important than ever for all students to have opportunities to learn CS. Traditionally, CS is not part of academic required subjects nor is it available in all U.S. K-12 schools [1]. With growing efforts and support for CS

© Springer International Publishing AG 2016
A. Brodnik and F. Tort (Eds.): ISSEP 2016, LNCS 9973, pp. 142–154, 2016.
DOI: 10.1007/978-3-319-46747-4_12

education, including districts, cities, and even states planning to implement CS across K-12 schools as well as a federal initiative announced in early 2016 called Computer Science for All, we conducted comprehensive U.S. research on CS at the K-12 level to evaluate the progression. We sought to understand the landscape and perceptions of CS for students across the U.S. in order to inform these efforts and to advance CS education at the K-12 level.

We focus on the landscape of access because positive correlations between computer use and attitudes towards computing have been well-documented [1, 2]. In fact, from the 1980s–90s, U.S. policies widely implemented technology in schools in an attempt to equalize educational opportunity [1, 3]. However, little was done to ensure these technologies were used effectively across demographics, even within schools. Lack of emphasis on advanced computing excluded CS advancement for those without the opportunity to learn CS otherwise. Further, home computer use was much lower for Blacks and Hispanics [4], while access was lower and later for girls [5]. We analyze a national sample to provide updates on which students have access and exposure to tech and CS, decades after the policies.

One major challenge in CS education is the lack of diversity. Specifically in the U.S., the lack of diversity includes the underrepresentation of women, Blacks, and Hispanics. At the high school level, Advanced Placement (AP) CS A participation is low overall, but drastically lower for Black and Hispanic students, comprising only 3.9 % and 8.8 % of test takers in 2014, respectively. At the university level, only 11.4 % of CS degrees were awarded to Blacks and 8.5 % awarded to Hispanics in 2012 [6]. And with the release of diversity data by top technology companies, the lack of diversity is sounding the alarm for action.

Perceptions, encouragement, and exposure play important roles in the lack of participation and interest [7]. Incorrect or lack of perceptions as well as stereotypes may discourage students from studying CS [8, 9]. Similarly, self-perceptions in one's own ability has been found to be correlated with interest and participation in STEM and CS [10–12].

External influences involve encouragement from adults as well as peers and increase the likelihood of pursuing and persisting in STEM and CS [10, 13, 14]. Adults, including teachers, can powerfully influence students [1]. Teachers' low expectations have negatively affected students' short- and long-term performance [15, 16]. Traditional roles and the "geek," male, and White stereotype may exclude many from feeling a belonging in CS [17] while influencers may disproportionately encourage certain types.

The lack of CS education exposure to students is possibly a large reason for the lack of knowledge and encouragement in the field [8]. Therefore, beyond perceptions and encouragement, we wanted to dig deeply to understand the landscape of CS access for students in the United States and investigate the interrelated factors behind the low numbers pursuing CS.

This paper details the first two years of a three-year study on the landscape of CS education for K-12 students in the U.S., surveying students, parents, teachers, principals, and superintendents. The goal of the study is to understand

- knowledge and perceptions of CS
- disparities in access, and
- barriers to offering CS.

2 Methodology

This study details results from the first two years of implementation, surveying 1,673 students, 1,685 parents, 1,013 teachers, 9,693 principals, and 1,865 superintendents in 2014 and 1,672 students, 1,677 parents, 1,008 teachers, 9,244 principals, and 2,227 superintendents in 2015–16, representative across the United States. Samples from the two years are not necessarily the same, though because individuals were polled from the same panels, there may have been overlap.

Telephone surveys were conducted with students, parents, and teachers currently living in all 50 states and the District of Columbia using a combination of two sample sources: the Gallup Panel and the Gallup Daily tracking survey. The Gallup Panel is a proprietary, probability-based panel of U.S. adults selected using random-digit-dial (RDD) and address-based sampling methods. The Gallup Panel is not an opt-in panel. The Gallup Daily tracking survey sample includes national adults with a minimum quota of 50 % cellphone respondents and 50 % landline respondents, with additional minimum quotas by time zone within region. Landline and cellphone numbers are selected using RDD methods. Landline respondents are chosen at random within each household based on which member had the most recent birthday. Eligible Gallup Daily tracking respondents who previously agreed to future contact were contacted to participate in this study.

Student and parent samples included targeted, detailed data on the underrepresented (Blacks and Hispanics, including Spanish-speaking only). Students were in grades 7–12 (around age 12–18) and parents had a child in grades 7–12. Teachers taught 1st–12th grade (around age 6–18), with approximately 21 % teaching or have taught computer science. The population for principals was sampled from a list of 99 % of U.S. public schools and approximately 30 % of U.S. private schools. The population for superintendents was from a panel including more than 20 % of all U.S. K-12 school district superintendents.

Student and parent samples were weighted to correct for unequal selection probability and nonresponse. Student data were weighted to match national demographics of age, gender, race, ethnicity and region. Parent data were weighted to match national demographics of age, gender, education, race, ethnicity and region. Demographic weighting targets were based on the most recent Current Population Survey. Teacher samples were weighted to correct for unequal selection probability and nonresponse. The data were weighted to match demographics of age, gender, education, race, ethnicity and region. Demographic weighting targets were based on the Gallup daily tracking information. Principal and superintendent samples were weighted to match national demographics of school ZIP code, school enrollment size, and census region.

Surveys for all five groups covered topics on perceptions of CS, interest in and desire for CS, in- and out-of-school opportunities for CS, participation in CS, and obstacles to providing and accessing CS opportunities. Survey items were closed-ended, with agreement for yes/no questions and Likert scales for agreement with statements

(1–3 Likert for students and parents and 1–5 Likert for teachers, principals, and superintendents). Surveys were not completely the same from the first year to second year, as new questions were introduced based on findings from the first year. Many questions were kept in order to track trends over the entire research study. The appendix includes sample questions.

The surveys for students, parents, and teachers each lasted about 10 min over the phone, with 30–40 questions. Principals and superintendents were emailed online surveys. Principal surveys had approximately 30 questions. Superintendents were surveyed as part of another regular online survey, with 10 closed-ended questions for this study.

After data were collected, a rigorous quality assurance process was used to clean the data. The data were then coded and reviewed by response. Indices of related variables were created and analyzed using regression to understand trends across and within the surveyed populations.

3 Findings

3.1 Knowledge and Perceptions

From the first-year survey, we found that most respondents do not have a clear understanding of what computer science is, and responses were varied. The misunderstanding was that CS includes creating documents and presentations (78 % of students, 64 % of parents, 75 % of teachers, and 63 % of principals said this) as well as searching the Internet (57 % of students, 49 % of parents, 64 % of teachers, 54 % of principals said this). In particular, Black or Hispanic students are somewhat less knowledgeable about computer science. Female students, parents, and teachers were also less knowledgeable about computer science. This confusion with basic computer literacy is important to distinguish, particularly for educators and parents who may believe they are providing CS opportunities.

However, after this initial question gauging understanding of CS, respondents were presented with a definition of CS (see Appendix). So once they understood CS, we see that perceptions of computer science are very positive and high across all populations. Most agree that people who do CS make things that help improve people's lives (93 % of students and parents, 86 % of teachers, 82 % of principals, and 76 % of superintendents agree) and that people who do CS work on fun and exciting projects (with 91 % of students and 94 % of parents agreeing in the first year, increasing to 94 % of both students and parents in the second year; and 82 % of teachers and 78 % of principals agree). Interestingly, we see that students and parents are most likely to have positive perceptions of computer science careers and work, whereas educators are less likely to have positive perceptions, with increasing authority (and distance from students) correlating with lesser positive perceptions.

Further, we see that all populations have high utility value [18] of computer science careers. Over 96 % of parents and students agree that CS can be used in a lot of different types of jobs, and 81–89 % of teachers, principals, and superintendents agree. And, most students (68 %) and parents (79 %) agreed that computer scientists have good-paying jobs. Over 86 % of students and parents say that the student will "somewhat likely" or "very likely" have a job where they need to know CS.

In addition, the majority of all groups support CS in schools and believe it is important. In the first year 90 % of parents said that they thought offering CS is a good use of school resources, which increased to 93 % in the second year. When comparing with required courses like math, science, history, and English, over 84 % of parents, 71 % of teachers, and two-thirds of principals and superintendents thought CS was more or just as important. Roughly 9 in 10 of the adults say the CS is just as or more important than elective courses like art, music, and foreign languages. Black or Hispanic parents are more likely than White parents to say that CS is even more important than the required or elective courses. Over half of teachers, principals, and superintendents think that most students should be required to take CS. With such high support, about 7 in 10 educators agree that it is a good idea to incorporate CS into other subjects at school.

Yet, while perceptions and value are very positive across populations, images of who does CS are very narrow. Half of students and parents agreed that you need to be very smart to do CS. Teachers (38 %) were less likely to agree and principals (19 %) and superintendents (17 %) were the least likely to agree with the statement. In terms of types of students, teachers (62 %) and principals (56 %) agreed that students good at math and science would be more successful in computer science. But only 42 % of students rated themselves as "very skilled" in math and 40 % as "very skilled" in science. Further, 56 % of students said they were "very confident" they could learn CS. Specifically, we saw that Hispanic students are less likely than White or Black students to say they are "very skilled" at science. Hispanic students are also less likely to say they are very confident they could learn computer science if they wanted to. With lower confidence, Hispanic students may be less likely to be encouraged or interested in fields like CS.

In the media, not surprisingly, both students and parents perceive those who do computer science as mostly White, male, and "wearing glasses." We also saw that of the students who said they saw people doing CS in the media, only 16 % said that they often see people who are like them. By gender, we see a stark contrast: 21 % of boys said they "often" see people like them while only 11 % of girls said "often." In fact, 31 % of girls said they "never" see someone like them while only 18 % of boys said "never." Thus, students who don't identify as looking like who they perceive as computer scientists or who don't identify as nerdy or smart may not feel a sense of belonging with computer science. With lower confidence and sense of belonging, certain students may be less likely to be encouraged, less interested, and less likely to learn CS, creating a cycle of reinforcement. Interestingly, by race, 13 % of Hispanic students, 16 % of White students, and 26 % of Black students said that they see someone like them doing CS "often" in the media. This points to other complex factors that may be at play. Overall, these findings imply a need to better shape CS learning environments and social influencers (from the media to educators to parents and to industry) to be inclusive of all backgrounds in order to diversify the students learning CS.

3.2 Disparities in Access

Many schools do not offer CS, with disparities by demographic. About 40 % of teachers and principals said that their school did not have any dedicated CS classes. These numbers improved from the first year survey to the second: 43 % of principals in the first year reported having no CS classes, which decreased to 39 % in the second year. Black students are less likely to report having access to CS classes and CS taught in other classes. Only about 1 in 5 of these principals said they offer Advanced Placement CS, an advanced course that allows students to receive university credit. However, the content is trending to more likely include programming and coding. In the first year, 53 % of principals reported that these CS opportunities included programming and coding, which increased to 66 % in the second year. In terms of other programs, only about half of teachers and principals said that their school offers CS groups or clubs, with numbers roughly the same in both years.

For opportunities outside of school, only about half of students and parents are aware of opportunities in their community to learn CS. And just slightly more, about two-thirds and 54 % of parents, are aware of specific websites to learn CS. Technically, online opportunities are available to anyone anywhere, so it is surprising that not more parents are aware of these websites. Male students in particular are more likely to be aware of opportunities in the community and online than female students. Parents of boys are also more likely to be aware of these opportunities. And, Hispanic students and parents are less likely to be aware of opportunities in the community. These discrepancies in awareness of CS opportunities fall in line with images of who does CS.

Exposure to technology also has disparities. Hispanic students are less likely to know an adult working in technology (49 % versus 68 % of Whites and 65 % of Blacks) and less likely to use a computer everyday at school (31 % versus 42 % of Whites and 34 % of Blacks) or at home (26 % versus 45 % of Whites and 30 % of Blacks, with 10 % of Hispanics saying they never use a computer at home).

In terms of how many students have learned CS, 53 % of students in the first year of the study said they learned CS, increasing to 55 % of students in the second year who said they learned CS. Boys are more likely to say they have learned CS than girls (59 % boys versus 50 % girls). Among students who stated learning CS, 73 % in the first year and 80 % in the second year said they learned it in a class at school and about half learn it on their own, outside of any group or program (56 % in the first year and 48 % in the second year). About a third learn it online through a class or community and about a quarter learn it through a group or club at school or through a group or program outside school. In particular, boys are more likely to have learned CS outside of school: online through a class or group, in an afterschool group or club, or on their own. Black and Hispanic students are also more likely to have learned CS in a group or club at school, and Black students are more likely to have learned it in a group or program outside of school.

When asked about computational thinking (CT) [19], only about 37 % of students said they've done CT at school while 68 % of teachers said that they've incorporated CT into their classes. More students reported doing specific CT activities. Thus, students may be learning CT without realizing that what they are doing is considered CT.

3.3   Barriers

Despite the positive value and high support of CS among parents, as discussed earlier, we saw that few principals and superintendents thought that demand from students and parents was high. Less than 8 % of principals and superintendents reported that demand from parents was high across both years of the study. We explored this further in the second year of the study and found that of parents and teachers who have expressed support for classes or curriculum to the school or principal, only one third of them have specifically expressed support for CS.

Consequently, we also saw that few educators believed that CS was a top priority at their school or district. Only about 1 in 5 teachers and principals agreed with this, and less than 30 % of superintendents agreed. Just over 40 % of principals reported that teachers and guidance counselors thought CS was important to offer and roughly the same percentage of teachers, principals, and superintendents believed that their school board was committed to offering CS. A large portion, about 25–40 %, also stated that they did not know or were neutral about teacher, guidance counselor, and school board support.

In both years of the study, we found that the greatest barriers to offering CS were related to lack of a qualified teacher, budget to train or hire a teacher, as well as the need to devote time towards standardized testing requirements[1] rather than computer science. Over 73 % of superintendents in both years said that a barrier to offering CS is they do not have teachers at the school qualified to teach CS. In the first year of the study, 42 % of principals said this, increasing to 63 % in the second year. Around 56 % of superintendents in both years said that there was not enough money to train or hire a teacher. This same barrier also became more prevalent for principals, increasing from 44 % who said this in the first year to 55 % in the second year. And about half of principals and superintendents stated that the need to devote time to testing requirements and CS is not a testing requirement, increasing slightly from 47 % of principals and 52 % of superintendents in the first year to 50 % of principals and 55 % of superintendents in the second year. Overall, the most common single "main reason" for not offering CS were the testing requirements, cited by about 30 % of principals and 23 % of superintendents.

Despite the lack of perceived demand, the lack of prioritization, and the challenges with obtaining qualified CS teachers and testing requirements, an opportunity lies in incorporating CS into existing subjects. As noted earlier, about 7 in 10 educators agree that it is a good idea to incorporate CS into other subjects at school. In the second year, we found that 29 % of teachers say they have already incorporated some elements of CS in their classes. And, 62 % of teachers reported that they know where to learn more about incorporating CS and 58 % said they would be willing to spend their own time to learn more about CS. Because elements of CS – programming/coding and CT – are tools of critical thinking, problem solving, and creative expression, teachers can and have effectively incorporated CS into various subjects in order to teach content knowledge by means of CS.

---

[1] In the U.S., public school students are required to take annual standardized tests in math, reading, and in later years science to provide a measure of how students and schools are performing.

4   Conclusion

With all the momentum of CS education in K-12 schools, there is still a need to distinguish CS from basic computer literacy among all populations, including educators, so that students are engaged in opportunities to advance beyond using computers to creating technologies and tools. And more work is also needed to dispel stereotypes of who does CS, even with the high value of CS and positive image of CS careers and work across all groups – students who don't fit these stereotypes often lack access to these CS opportunities and are even unaware of existing opportunities. Further, with barriers like discrepancies in perceived demand for CS, lack of prioritization, obtaining qualified teachers, and testing requirements, the education infrastructure does not provide all students with the needed exposure, particularly Blacks and Hispanics. We also saw that girls are less likely to have learned CS. Finally, to overcome challenges in the education infrastructure, we saw an opportunity to incorporate CS into existing subjects. Our findings suggest:

- We need to increase awareness of the differences between basic computer literacy and CS;
- Computer science training resources that are inclusive of all students need to be made accessible, available, and known;
- Influencers should be aware of the images they promote and diversify images of those who do CS;
- Educators should talk to parents, and parents should speak up about their demands in CS;
- Policymakers and administrators should consider strategies to be more supportive of K-12 computer science offerings, such as
    - more flexible curriculum and class schedules,
    - modifying requirements for standardized testing,
    - allowing computer science courses to count towards graduation and college admission requirements,
    - offering a variety of paths to learn computer science in and out of school, as well as through various means using computers, mobile devices, and without technology.

Acknowledgements. We thank the Gallup team for their partnership, including Katherine Black, Cynthia English, Elizabeth Keating, Brandon Busteed, Stephanie Kafka, Dawn Royal, and countless others. We would also like to thank the many individuals at Google and in the CS education community who have supported us from developing survey items to reviewing drafts, including Chris Stephenson, Jason Ravitz, Mo Fong, and many more.

Appendix

Sample survey questions. For agreement statements, students and parents were given 1–3 Likert scale and teachers, principals, and superintendents were given a 1–5 Likert scale.

Knowledge of CS. Based on what you have seen or heard, which of the following activities do you consider part of computer science? (yes, no, don't know, refused).

- Programming and coding
- Creating new software
- Creating documents or presentations on the computer
- Searching the Internet

After this first question (only for students, parents, teachers, and principals), respondents were provided a definition of CS and reminded multiple times throughout the survey:

> *Computer science can involve MANY types of activities. Today we are only going to focus on a specific type of computer science.*
> *For the purposes of this survey, computer science is the study of how computers are designed and how to write step-by-step instructions to get them to do what you want them to do. This is sometimes referred to as computer programming or coding. Computer science includes things like creating software, applications, games, websites and electronics and managing large databases of information.*
> *For the purposes of this survey, computer science does NOT include using a computer to do everyday things, such as browsing the Internet. Please keep this definition in mind as you answer the following questions.*

Images of CS

- People who do computer science make things that help improve people's lives.
- People who do computer science have the opportunity to work on fun and exciting projects.
- Computer science can be used in a lot of different types of jobs.
- Most people who work in computer science have good-paying jobs.
- Students who are good at math and science are much more likely to succeed in learning computer science.
- People who do computer science need to be very smart.

Self-image

- How confident are you that you could learn computer science if you wanted to? Very confident, somewhat confident, or not very confident?
- How likely are you to have a job someday where you would need to know some computer science? Is it very likely, somewhat likely, or not at all likely?
- How often do you see people who do computer science in movies or TV shows who are (read and rotate Q04A–Q04F)? Do you see them most of the time, some of the time, not very often, or never?

  - Women
  - White
  - Black or African-American
  - Hispanic/Latino
  - Asian
  - Wearing glasses
- How often do you see or read about people doing computer science in each of the following places? In TV shows (Often, Sometimes, Never)
- How often do you see or read about people doing computer science in each of the following places? In movies (Often, Sometimes, Never)
- How often do you see or read about people doing computer science in each of the following places? Online through social media, articles, or videos (Often, Sometimes, Never)
- Thinking about all of the people you see or read about doing computer science in TV shows, in movies, or online, how often do you see people like you doing computer science? (Asked only of those who see people doing CS "OFTEN" or "SOMETIMES" on TV, movies, and/or online) (Often, Sometimes, Never)

Exposure to Technology

- How often do you use a computer at your school? (Every school day, Most school days, Some school days, Never)
- In a typical week, how often do you/does your child use a computer at HOME? (Every day, Most days, Some days, Not very often, Never)
- In a typical day, how many hours do you/does your child use a computer at HOME? (Asked only of students/parents who use a computer with Internet at home every day) (Less than 2 h, 2–5 h, More than 5 h)
- In a typical week, how often do you/does your child use a cell phone or tablet? (Every day, Most days, Some days, Not very often, Never)
- In a typical day, how many hours do you/does your child use a cell phone or tablet? (Asked only of students/parents who use a cell phone or tablet every day) (Less than 2 h, 2–5 h, More than 5 h)

Learning CS

- Have you ever learned computer science in any of the following ways? (yes, no, don't know)
  - In a class at school
  - In a group or club at school
  - In a formal group or program outside of school, such as a camp or summer program
  - Online through a class, program, or online community
  - On your own outside of a class or program

Value of CS

- It is a good idea to try to incorporate computer science education into other subjects at school.

- Offering opportunities to learn computer science is a good use of resources at your child's school.
- Do you think offering opportunities to learn computer science is more important, just as important, or less important to a student's future success than required courses like math, science, history and English?
- Do you think offering opportunities to learn computer science is more important, just as important, or less important to a student's future success than other elective courses like art, music, and foreign languages?
- Most students should be required to take a computer science course.

Demand

- Which of the following best describes the demand for computer science education among parents in your school/district? Is demand... (high, moderate, low)
- Which of the following best describes the demand for computer science education among students in your school/district? Is demand... (high, moderate, low)

Priority

- My school board believes computer science education is important to offer in our schools.
- Computer science education is currently a top priority for my school/district.
- The majority of teachers at my school think it is important to offer opportunities to learn computer science.
- The majority of guidance counselors at my school think it is important to offer opportunities to learn computer science.

Barriers

- As far as you know, why doesn't your school offer any ways to learn computer science? (check all that apply)
  - There are no teachers available at my school with the necessary skills to teach computer science.
  - There are no teachers available to hire with the necessary skills to teach computer science.
  - There is not enough classroom space.
  - There is not enough money to train or hire a teacher.
  - We do not have the necessary computer equipment.
  - We do not have the necessary computer software.
  - We do not have sufficient budget to purchase the necessary computer equipment.
  - We do not have sufficient budget to purchase the necessary computer software.
  - Internet connectivity is poor at my school.
  - There is not enough demand from students.
  - There is not enough demand from parents.
  - There are too many other courses that students have to take in order to prepare for college.

- – We have to devote most of our time to other courses that are related to testing requirements and computer science is not one of them.
- – Don't know
- Among the reasons just mentioned, what would you say is the MAIN reason your school doesn't offer ways to learn computer science?

References

1. Barker, L.J., Aspray, W.: The state of research on girls and IT (2006). https://lexus.ischool. utexas.edu/Westbrook_Lynn/2008/fall/INF180J/aspray_stateofresearch.pdf
2. Levine, T., Donitsa-Schmidt, S.: Computer use, confidence, attitudes, and knowledge: a causal analysis. Comput. Hum. Behav. **14**(1), 125–146 (1998)
3. Margolis, J., Estrella, R., Goode, J., Holme, J.J., Nao, K.: Stuck in the Shallow End: Education, Race, and Computing. MIT Press, Cambridge (2010)
4. Jesse, J.K.: The digital divide: political myth or political reality? In: Aspray, W. (ed.) Chasing Moore's law: Information Technology Policy in the United States. SciTech Publishing, Raleigh (2004)
5. Kirkpatrick, H., Cuban, L.: Should we be worried? What the research says about gender differences in access, use, attitudes, and achievement with computers. Educ. Technol. **38**(4), 56–61 (1998)
6. National Center for Education Statistics (NCES): Degrees in computer and information sciences conferred by degree-granting institutions, by level of degree and sex of student: 1970–71 through 2010–11 (2012). http://nces.ed.gov/programs/digest/d12/tables/dt12_349. asp
7. Wang, J., Hong, H., Ravitz, J., Ivory, M.: Gender differences in factors influencing pursuit of computer science and related fields. In: Proceedings of the 2015 ACM Conference on ITICSE, pp. 117–122. ACM (2015)
8. Carter, L.: Why students with an apparent aptitude for computer science don't choose to major in computer science. ACM SIGCSE Bull. **38**(1), 27–31 (2006)
9. Cheryan, S., Plaut, V.C., Handron, C., Hudson, L.: The stereotypical computer scientist: gendered media representations as a barrier to inclusion for women. Sex Roles **69**(1–2), 58–71 (2013)
10. Denner, J., Werner, L., Martinez, J., Bean, S.: Computing goals, values, and expectations: results from an after-school program for girls. J. Women Minor. Sci. Eng. **18**(3), 199–213 (2012)
11. Cheong, Y.F., Pajares, F., Oberman, P.S.: Motivation and academic help-seeking in high school computer science. Comput. Sci. Educ. **14**(1), 3–19 (2004)
12. Smith, T.J., Pasero, S.L., McKenna, C.M.: Gender effects on student attitude toward science. Bull. Sci. Technol. Soc. **34**(1–2), 7–12 (2014)
13. Fan, X., Chen, M.: Parental involvement and students' academic achievement: a meta-analysis (2001). http://rd.springer.com/article/10.1023/A:1009048817385
14. Dabney, K.P., Chakraverty, D., Tai, R.H.: The association of family influence and initial interest in science. Sci. Educ. **97**(3), 395–409 (2013)
15. Lavy, V., Sand, E.: On the origins of gender human capital gaps: short and long term consequences of teachers' stereotypical biases (No. w20909). National Bureau of Economic Research (2015)
16. Rosenthal, R., Jacobson, L.: Pygmalion in the classroom. Urban Rev. **3**(1), 16–20 (1968)

17. Cheryan, S., Master, A., Meltzoff, A.N.: Cultural stereotypes as gatekeepers: increasing girls' interest in computer science and engineering by diversifying stereotypes. Front. Psychol. **6**, 49 (2015)
18. Master, A., Cheryan, S., Meltzoff, A.: Computing whether she belongs: stereotypes undermine girls' interest and sense of belonging in computer science. J. Educ. Psychol. (2015). http://dx.doi.org/10.1037/edu0000061
19. Wing, J.M.: Computational thinking. Commun. ACM **49**(3), 33–35 (2006)

Combining the Power of Python with the Simplicity of Logo for a Sustainable Computer Science Education

Juraj Hromkovič, Tobias Kohn, Dennis Komm$^{(\boxtimes)}$, and Giovanni Serafini

Department of Computer Science, ETH Zürich, Universitätstrasse 6,
8092 Zürich, Switzerland
{juraj.hromkovic,tobias.kohn,dennis.komm,giovanni.serafini}@inf.ethz.ch

Abstract. Computer science education in K-12 and for non-majors at university often aims at making students confident with computational thinking by introducing them to programming. We are convinced that such programming classes offer a great opportunity to expose students to core concepts of computer science and thereby contribute to a broad and general education.

In this article, we describe our approach and experiences with teaching programming at various levels, namely at primary schools, high schools, and universities. We identify a set of goals that allow us to go beyond the pure teaching of specifics of a given programming language, i. e., syntactical details, and shift the focus towards sustainable topics such as algorithms as problem solving methods and their analysis.

1 Introduction

Programming education is not primarily about teaching a specific programming language or using a computer. It is about introducing the language and thinking of computer science itself. In this context, programming education then unfolds merits for a broad and general education and acts as the doorway not only into computer science.

This article describes our efforts, directed towards establishing programming and computer science education on all three levels of education: primary school, high school, and university. With each level having its own specific focus and adapted curriculum, they eventually lead to a set of common goals. Bringing programming education to schools, however, cannot be limited to the design of curricula. We are actively involved in reaching out to teachers at primary and high schools, and in training them to successfully teach programming and computer science.

Our classes are all taught in Logo and Python, respectively. Both these languages serve our goals very well, particularly in that they allow a focus on concepts rather than the language itself. This way, concepts such as the modular design of algorithms can be taught without a lengthy introduction to syntactical details. And we can start with simple commands that directly induce a visual

A. Brodnik and F. Tort (Eds.): ISSEP 2016, LNCS 9973, pp. 155–166, 2016.
DOI: 10.1007/978-3-319-46747-4_13

feedback (e. g., by moving the turtle on the screen). This way, testing and correcting programs becomes available already for beginners and children. After that, the students can write their first programs and understand that, essentially, these programs increase the vocabulary of the computer. Building larger programs that use smaller ones as building blocks is nothing else than using a modular design to solve tasks of increasing complexity.

Neither Logo nor its turtle graphics are limited to drawing figures. As an example, Logo's simple `repeat`-loop allows for an easy way to understand the concept of loops without introducing variables or conditionals. Even the turtle itself has proven to be an invaluable asset beyond graphics due to the rich metaphors it provides.

Organization of This Paper
Section 2 gives an overview of the context of our programming education, our goals, and our motivation. Sections 3, 4 and 5 then discuss the respective implementation on the three levels of education, i. e., primary school (Sect. 3), high school (Sect. 4), and university (Sect. 5). Section 6 contains concluding remarks.

2   Setting, Goals, and Motivation

In this section, we describe the Swiss educational system, and the common goals we aim to achieve with our approach. In particular, we identify the learning goals, which are independent of the concrete choice of the programming language.

2.1   Swiss Educational System

Switzerland is a pronounced federalist country comprising 26 states and embracing the four official languages German, French, Italian as well as Romansh. The Swiss education system reflects the political organization of the country and has to take care of the regional specificities of the population. Therefore, the responsibility for education historically lies with the states, each of them having its own dedicated minister.

The compulsory education takes 11 years and encompasses kindergarten, primary school as well as lower secondary school. This compulsory part is followed by four years of high school and eventually university, or a practical apprenticeship with accompanying schools.

Current efforts to constitute a common standardized curriculum for primary and lower secondary schools include computer science as a mandatory subject. With the programming courses we have conducted in the last few years, we have already shown a viable way of introducing students to computer science.

2.2   Learning Goals

Algorithms are the scientific core of computer science and confer the discipline its conceptual identity. Knowledge about algorithms is nowadays an indispensable asset in society, industry, and research. Algorithms should therefore play an adequate role in school education. It is important to note that the term "algorithm" is not bound to a concrete programming language. It is the formal finite description of a method that solves all (usually infinitely many) instances of a given (computational) problem within finite time. A scientifically sound computer science subject at school should focus on the study of algorithms, from a theoretical as well as from a practical perspective. In addition, a spiral curriculum through the different stages of education allows to discuss different aspects of algorithms adequately at the appropriate level.

Out of the many aspects of algorithmic and computational thinking, we have identified three aspects as the primary goal of programming education: introducing the programming language as an example of a formal language, understanding programming as the process of automation and abstraction, and discussing the limits of practical computability.

The Concept of a Formal Language. Students start out with a very limited set of words, each standing for a specific instruction given to the machine. A program is then a sentence of such words and must follow certain rules imposed on grammar and syntax. As the students progress in the curriculum, they expand their vocabulary not only through acquisition of instructions but more importantly through the creation of new words of their own. They learn that a given instruction set does not suffice to solve all problems in an elegant way, but can and has to be expanded.

Formal languages clearly are the building blocks of both the human computer interface and the algorithms used in the sciences. Moreover, the notion of extending and adapting a formal language to specific needs is key in dealing with complex problems and formulating intelligent solutions.

Automation and Abstraction. As the students' programs grow, it becomes infeasible to spell out every detail. Students must rely on principles of automation and abstraction, and employ looping constructs and subroutines in order to manage otherwise unwieldy programs. We use a bottom-up approach to modularization here, starting with simple programs, which then continuously grow in size and complexity.

In recent years, the question of automation has attracted increased attention in the form of big data problems. Concepts of how to manage huge data sets and automate computations and manipulations on that data go far beyond computer science and have entered political debate on more than one issue.

Limits of Practical Computability. Students are exposed to questions of the limits of computability from the beginning on. Lacking precision quickly becomes an issue when dealing with floating point numbers in the context of graphics, say. Approximations are inevitable and students have to learn to take (numerical) errors into consideration when designing their programs.

Besides the available numeric precision, students must also deal with the efficiency of their programs and time constraints. Measuring (or even roughly analyzing) the execution time of a program is the first step towards a more general discussion of the running time of algorithms. Eventually, this discussion will include topics such as polynomial versus exponential running time.

Computers do not only offer exciting new possibilities. There are also some clear cut limitations on what a computer can achieve, both in precision and extent. The study of this feasibility sets computer science apart from pure mathematics and is the motivation of many advancements in the field of algorithms. Most prominently, modern cryptography rests on principles of intractability in the sense of computer science, even though these problems are solvable when seen from the point of mathematics.

2.3   Turtle Graphics

Programming always takes place in the context of a specific machine model that executes the code and provides a set of available instructions. Novice programmers first have to understand the basic properties of this model or *notional machine* they are going to program [4].

Turtle graphics provides an excellent model of a programmable machine. The current state and the properties of the (virtual) turtle are directly observable and the basic instructions for movement fit well into the student's mental models. Furthermore, the turtle serves as a metaphor for introducing various new concepts. Defining a new function, for instance, can be motivated as "teaching the Turtle a new word" [11]. Yet, the students learn that the communication with the computer (the turtle, respectively) needs to be precise. Since computers have no intellect, there is no room for interpretation.

The second important aspect of turtle graphics is its immediate and direct feedback to the student. Mistakes in a program show immediately, and our experience confirms that students are highly motivated to correct their programs until the desired output is achieved. In this way, the important concepts of testing, verifying, and correcting programs can be introduced immediately while starting to write first programs. Writing correct programs without the aid of turtle graphics is much more abstract and students tend to think of correct programs as those which compile and execute.

As students progress in the curriculum, the actual turtle becomes less important and only the result of the drawing process is shown. It turned out that, in this mode, turtle graphics is fast enough to even support simple animations, games, and mathematical visualizations.

3   Primary Schools

The students we introduce to programming are usually 10 to 12 years old and attend the fifth or the sixth grade of primary school. A programming course comprises twenty 45-minutes units, which are split up into blocks of either two or four units a week. The students are taught by a team comprehending a lecturer, an assistant as well as the regular class teacher. The responsibility for the class lies with the lecturer, while the assistant and the class teacher are expected to individually support students who need specific advice.

We embed the classes into the usual school activities on-site at primary school. Swiss schools are allowed to autonomously organize ad hoc activities for their students and can therefore reserve time slots for projects like ours. Our programming classes are mostly part of regular math lessons [12].

The Logo programming courses at primary schools started more than 10 years ago as an initiative of our chair at ETH Zurich. The primary objectives consisted in developing and continuously improving teaching materials for students, in holding classes directly at school, in educating class teachers without prior knowledge in computer science, and in practically demonstrating how a didactically and pedagogically adequate but scientifically sound introduction to programming at primary schools can be implemented.

The requests from schools and teachers rapidly rose. At the beginning of 2014, we launched the so called PrimaLogo project and thus started to spread the courses all over the country. During the school year 2015/16, we and our regional partners visited more than 80 school classes and introduced roughly 1600 students and their teachers to programming with Logo.

3.1 Settings and Goals

The goal of the programming courses is to teach the students how to interact with the computer using a programming language. With respect to the common goals introduced in Sect. 2.2, we focus here on the concept of a formal language as well as on automation and abstraction. The students are taught that the turtle they have to move on the screen has a mother language (Logo) and a very restricted vocabulary. Each of the words in this vocabulary corresponds to one instruction the turtle can unambiguously understand and execute.

The students learn that a program is a sequence of words taken from the available vocabulary and understand that the abstract activity of programming consists in writing sentences in the mother language of the turtle. Furthermore, the students gradually realize that the initial vocabulary may not be sufficient to meet all the expectations they have in the turtle, and learn that a new word can be used to stand for a sentence they already wrote. The students therefore learn how to extend the language of the turtle by giving their programs a name.

We rely on the programming language Logo and believe that Logo still is one of the most adequate programming languages for novices, particularly for classes at primary school. Our didactic approach focusses on a small subset of the programming language and aims to permit the students to care for correct syntax as well as to avoid the cognitive overload caused by an unmanageable list of instructions or by an overcharged graphical user interface [12,13]. The chosen programming environment is XLogo4Schools [14], a redesigned version of the open source application XLogo [10].

Logo allows to teach structured programming and modular development as a key problem solving pattern in a natural and didactically effective way. For instance, as already mentioned, Logo's `repeat`-loop does not require the students to deal with the abstract concept of a variable, and therefore allows to introduce this control structure already at the beginning of a programming class.

3.2 The Role of the Class Teacher

One of the primary objectives of our programming courses is to introduce class teachers to programming and its didactics. We aim to make class teachers confident with contents, teaching materials, and the didactic approach, so that they feel ready to autonomously employ the same course in their future classes.

In Switzerland, a primary school teacher is required to attend high school and to obtain a Bachelor of Arts from a school of education. Since, unfortunately, computer science plays no or a negligible role in the current curriculum at both these education levels, primary school teachers have no formal education in computer science and thus no prior knowledge in programming at the beginning of our projects at their schools.

To pragmatically yet adequately support class teachers, we focus on dedicated activities divided into three parts.

– *Theoretical Part.* The class teachers attend a half-day workshop in which they are introduced to programming relying on the same teaching materials and the same approach adopted in the programming course. Furthermore, lecturers and class teachers reflect on the contribution of computer science to education, on the role of programming classes, and on related didactic challenges.
– *Practical Part.* The class teachers support the guest lecturer and her or his assistant during the programming course in their class. For the class teachers, the programming course is a kind of safe, practical session, in which they actively help the students, but are not taking the responsibility for the lesson. Nevertheless, we encourage teachers to take over at least one activity with their class during the project and support them while preparing and carrying it out.
– *Individual Part.* In addition to the workshop and to the assistance during the courses, the class teachers are required to prepare the classes by studying the teaching materials as well as by solving the exercises on an individual basis.

3.3 Structure and Contents of the Teaching Materials

The school projects rely on the German textbook *An introduction to programming in Logo* (German: *Einführung in die Programmierung mit Logo*) [7], and on a Logo booklet [5] covering the contents of its first seven chapters. The chapters of the textbook include (1) programs as sequences of instructions, (2) simple loops with the `repeat`-instruction, (3) naming and calling programs, (4) regular polygons and circles, (5) programming animations, (6) programs with parameters, (7) passing parameters to subprograms, (8) how to optimize the length of a program and its computational complexity, (9) the concept of variables and the instruction `make`, (10) local and global variables, (11) branch instructions and `while`-loops, (12) integrating Logo and mathematics: geometry and equations,

(13) recursion, (14) integrating Logo and mathematics: trigonometry, and (15) integrating Logo and mathematics: vector geometry.

The textbook exemplifies how novices can be introduced to programming, embedding the key concepts into a spiral curriculum that starts at primary school and continues at lower and later at higher secondary school. To this end, the book describes the topics in such detailed way that students of secondary schools are even able to learn them individually, self-paced, and without a teacher.

Primary school students may be unable to cope with the form and the elaborateness of the textbook. To address this problem, the booklet mentioned above was developed. Here, the explanations are drastically reduced and the focus is on a simple language and on a comprehensive set of exercises. The students are expected to solve the exercises autonomously while the teacher has to accurately introduce them to the new concepts they are going to learn. The booklet is available online and free of charge for schools, students, and educators [5]. So far, it has been translated into English [6], French, Italian, Slovak, Serbian, Spanish, and Portuguese.

4 High Schools

At high school level, we find a wide range of approaches to teaching programming with little consensus on goals, contents, or methods. As outlined above, our approach is based on algorithmic and computational thinking. We use Python, extended by the loop structure taken from Logo, as explained further below.

4.1 The Setting

Due to its federal nature, the implementation of computer science education in Swiss high schools differs highly between schools. The mandatory part ranges from an introduction to current office applications to classes in programming and algorithmic thinking. Most schools also offer an advanced elective class in computer science for senior students.

One of the authors teaches at a school that offers an introduction to programming as part of an elective course in physics and applied mathematics. The course comprises two semesters in tenth grade with two classes each week.

Most students do not have any previous exposure to programming or computer science. So far, none of our high school students have previously attended a Logo course in primary school. Those students who bring prior programming knowledge are usually self-taught and lack a deeper understanding of algorithmic thinking.

4.2 Training Teachers

Teachers for programming classes in high schools have very diverse backgrounds. Most teachers are trained in mathematics or science but very few have an actual training in programming or even computer science.

As a result, one part of our efforts is concentrated on workshops open to all high school teachers. The *Annual National Day of Computer Science Education* (German: *Schweizer Tag für den Informatikunterricht*) is a prominent platform for brief informative workshops and serves to reach out to computer science teachers throughout the country [1]. More involved workshops of one or two days are then held independently and allow us to go deeper and discuss topics in more detail.

For a second part, we are involved in the training of future teachers that leads to a diploma in secondary and higher education from our university. Our lectures are usually attended by students of both mathematics and computer science. However, in contrast to the workshops, we concentrate rather on the education part and require from attending students to already bring a strong background in computer science.

4.3   Teaching Materials

The programming class is centered around a script, which has gone through a couple of iterations, taking into account experience from earlier classes. Out of the many lessons learned, the following five principles have guided the writing of the current script [8].

- Repeat each programming concept in various contexts to build a *spiral curriculum*. Instead of a single chapter about all looping constructs, each chapter contains a specific usage of loops and expands previous applications.
- *Short units* with as little explanatory text as possible, and with an emphasis on *hands-on* exercises. We found that most students have a strong tendency not to read but rather skim explanatory text.
- Complete and concise *example programs* to show how different elements are used and interact. A study by Lahtinen et al. has confirmed that example programs are the single most helpful teaching material [9].
- Put the focus on *plans and applications* of individual programming structures. For instance, the important (and difficult) concept is not a `for`-loop itself but how to use a `for`-loop to, e. g., compute a sum [4].
- Include additional problems for *fast students*. The heterogeneity of the classes has to be met by offering the faster students challenging problems and supplementary material.

According to these principles, the chapters are divided into short sections, with each taking two pages with a very brief introduction, an explained example program, and exercises, including supplementary problems for fast students. While the first chapter acts as a gentle introduction, corresponding to the Logo booklet used in primary schools, later chapters take into account that the class is part of a course in applied mathematics. In contrast to primary school level, we also include the aspect of the limits of practical computability from the beginning on (cf. Sect. 2.2). The chapters' contents are outlined in the following paragraphs.

(1) Learning the Basics with the Turtle. Students learn to use and control the turtle. Their first programming concepts are *user-defined functions* without return values (procedures), *parameters*, and *loops* with a fixed number of iterations. This first chapter of the script corresponds largely to the Logo booklet as explained in Sect. 3.3.

(2) Computations and Variables. Students learn to use Python without a turtle to perform computations and implement some simple algorithms. *Variables* are first introduced in the context of user input, together with *conditional execution* and motivated by the need for handling values not yet known at the time of programming. Later on, the concept of variables is then expanded to include changing values inside a loop.

(3) Coordinate Graphics. The turtle's screen is equipped with a coordinate system that lets the student direct the turtle to a specific point and even react to mouse events. In accordance with the idea of a spiral curriculum, there are no new programming structures introduced in this chapter. However, loops and conditional execution are both expanded with a discussion of nested structures.

(4) Functions. The true notion of *functions with return values* is introduced. Students need to learn how to define their own functions as well as how to use them properly. Our experience so far suggests that return values are the hardest concept of the entire curriculum. Most students express distinct difficulties with the correct use of the `return`-statement. The relationship between returning a result and printing a result on the screen seems to be particularly problematic.

(5) Lists. Students learn to use *lists* as a means to handle a (relatively) large amount of data. The main concept is to iterate through a list of data and perform operations such as searching, filtering, or sorting. This chapter does not make use of indexed access to list elements.

4.4 Logo's Loop in Python

Python is well-suited for classroom use. For our curriculum, however, we found one piece missing: a variable-free looping construct as in Logo. We have therefore extended Python by introducing such a looping construct and have gained very positive experience in that our students take loops as a natural concept and exhibit less problems writing programs with loops.

The very basic application of a loop is to have the computer repeat a part of the code for a given number of times. This might be a turtle drawing a regular polygon, say. Python provides two possibilities to achieve this task, using either a `for`- or a `while`-loop. While these are proven constructs, they both come with a penalty from an educational point of view: they use variables.

As educators we can either choose to introduce variables prior to loops, leave the occurring variables as a bit of magic in the code, or introduce loops together with variables at the same time. All three approaches are unsatisfactory. The first contrasts with the desire to use loops as early as possible in the curriculum as one of the starting points of abstraction. The second one comes with the danger of leaving students with the feeling of not being able to fully understand the programs they are to write. Finally, the third one comes with a steep learning curve as the students have to master at least two concepts simultaneously.

With Logo as model, we introduced a new looping construct to our Python-environment. The `repeat`-loop takes a number and then directly iterates the following code block without the need of a variable. Listing 1.1 shows a short sequence that uses our `repeat`-loop to draw a square.

Listing 1.1. Drawing a square using a simple `repeat`-loop.

```
repeat 4:
    forward(100)
    left(90)
```

5   Universities

We also use Logo at university level. Mainly, we introduce students of natural sciences to programming. In such an environment, the students have no background in computer science nor a particular interest in learning how to program a computer.

Following our discussion from Sect. 2, the first units of the lecture give a general introduction of how we communicate with computers in order to have them execute our programs and algorithms. It is pointed out that this needs to be done in an exact and unambiguous way since the machine does not have any intellect. As an example, we introduce the task of navigating the Logo turtle on the screen. The basic commands of Logo are described in the lecture and examples are given following the Logo booklet [5] (see Sect. 3.3). These basics (i. e., the main part of the booklet) are covered in roughly the first unit of the lecture. Next, the students are supplied advanced material that is specifically designed for this class [2]. One of the main topics introduced are variables, which are essential to subsequent units. Moreover, a simple `repeat`-loop is not sufficient anymore at this point, and the students are introduced to the concepts of *conditional execution* and the `while`-loop. Finally, they are given three *projects*, which ask to consolidate what they have learned so far by designing small programs to solve specific tasks [3]. The lecture is accompanied by exercise classes in which the students are asked to present and explain their solutions to a tutor.

After this introduction to programming with Logo, we continue with Python making use of its facilities to handle a wide range of data and scientific problems. Here, we implement projects that are related to the typical needs of, e. g., a biologist. However, the first lessons learned from Logo proved to be extremely valuable for the students, and they were able to apply many methods and paradigms to more complex algorithms.

6   Conclusion

Computer science with its algorithmic thinking has become an essential part of modern science. This prominent and important role renders it a necessity to include computer science in general education and thereby provide students of all levels with its rich set of tools for problem solving.

We envision a broad and comprehensive computer science education that starts in primary school and then continues by carefully building upon existing knowledge through secondary school, high school, and finally university. Starting point of such a curriculum could be the introduction to programming with Logo. The goal, however, is not the mastery of Logo as a specific programming language but rather the familiarity with algorithmic thinking from the beginning on.

Our experience with bringing Logo to primary schools has been overwhelmingly positive. Building upon that experience we included most of the Logo curriculum in more concise form at high school and university level, using Python, however, to tackle more abstract and advanced problems. Future work will further increase our collaboration between different levels and help establish computer science education as a central part of general education.

Acknowledgement. PrimaLogo is a cooperation of our chair, the Hasler Foundation, the Swiss Computer Science Teacher Association, the University of Basel, and the Universities of Teacher Education of Lucerne and of Graubünden. We are deeply grateful to all our project partners, the schools, the teachers, the local political authorities, the dozens of university students who teach and assist at school, and to the young school students for their contribution to the success of our activities.

References

1. Schweizer Tag für den Informatikunterricht (STIU) (2016). http://www.abz.inf. ethz.ch/stiu-2016-am-7-september-2016/
2. Böckenhauer, H.-J., Hromkovič, J., Komm, D.: Programmieren mit LOGO für Fortgeschrittene. http://abz.inf.ethz.ch/wp-content/uploads/unterrichtsmateri alien/primarschulen/logo_heft_2_de.pdf
3. Böckenhauer, H.-J., Hromkovič, J., Komm, D.: Programmieren mit LOGO – Projekte. http://abz.inf.ethz.ch/wp-content/uploads/unterrichtsmaterialien/primarsch ulen/logo_projekte.pdf
4. Boulay, B.D.: Some difficulties of learning to program. J. Educ. Comput. Res. **2**, 57–73 (1986)
5. Gebauer, H., Hromkovič, J., Keller, L., Kosírová, I., Serafini, G., Steffen, B.: Programmieren mit LOGO. http://abz.inf.ethz.ch/wp-content/uploads/unterricht smaterialien/primarschulen/logo_heft_de.pdf
6. Gebauer, H., Hromkovič, J., Keller, L., Kosírová, I., Serafini, G., Steffen, B.: Programming in LOGO. http://abz.inf.ethz.ch/wp-content/uploads/unterricht smaterialien/primarschulen/logo_heft_en.pdf
7. Hromkovič, J.: Einführung in die Programmierung mit LOGO - Lehrbuch für Unterricht und Selbststudium, 3rd edn. Springer, Heidelberg (2014)
8. Kohn, T.: Python. Eine Einführung in die Computer-Programmierung. http:// jython.tobiaskohn.ch/PythonScript.pdf

9. Lahtinen, E., Ala-Mutka, K., Järvinen, H.-M.: A study of the difficulties of novice programmers. In: Proceedings of the 10th Annual SIGCSE Conference on Innovation and Technology in Computer Science Education (ITiCSE 2005), pp. 14–18 (2005)
10. Loïc Le Coq:xLogo. http://xlogo.tuxfamily.org/. Accessed 28 Apr 2016
11. Papert, S.: Mindstorms.Basic Books, 2nd edn. (1993)
12. Serafini, G.: Teaching programming at primary schools: visions, experiences, and long-term research prospects. In: Kalaš, I., Mittermeir, R.T. (eds.) ISSEP 2011. LNCS, vol. 7013, pp. 143–154. Springer, Heidelberg (2011)
13. Sweller, J.: Cognitive load theory. In: Psychology of Learning and Motivation, vol. 55, pp. 37–76. Academic Press (2011)
14. Zivković, M.: Xlogo4school. http://sourceforge.net/projects/xlogo4schools/. Accessed 28 Apr 2016

A New Interactive Computer Science Textbook in Slovenia

Nataša Mori[1(✉)] and Matija Lokar[2]

[1] Faculty of Computer and Information Science, University of Ljubljana,
Ljubljana, Slovenia
natasa.mori@fri.uni-lj.si
[2] Faculty of Mathematics and Physics, University of Ljubljana, Ljubljana, Slovenia
matija.lokar@fmf.uni-lj.si

Abstract. Informatics is only a mandatory course in the first year and elective in the remaining three years in Slovene general secondary schools (grades 9–12). The course curriculum lists 100 learning objectives for all four years, but it does not specify the ones for the first (mandatory) year. Although this gives Computer Science teachers the freedom to choose the topics to be covered, they usually choose the ones covering digital literacy. One of the reasons is that this is the most important topic in the eighteen-year-old textbook that is currently used to teach Computer Science. In this paper we discuss a new interactive textbook that was introduced and the areas of knowledge it covers. We also discuss e-textbooks as a technology in general and give some feedback from teachers after the first year of using the new textbook.

1 Introduction

For the last 18 years Computer Science (CS) in Slovene secondary schools (grades 9–12) has not changed much. We have a mandatory first year of Computer Science for all pupils, the next three years are optional, as well as the Matura exam in Informatics. The paper talks about the four-year-long Computer Science subject as a whole, because in Slovenia, we have a single curriculum. The curriculum represents the topics and learning objectives (100 of them!) of the Computer Science subject, spread over four years. This means that teachers can be very flexible at preparing each year's course. The curriculum has pretty much stayed the same since 1998, with minor updates made in 2008. Updating the curriculum would definitely be the proper step towards better CS, but in our case, it is not possible. Fortunately, the committee of National Examination Centre followed the ACM trend in improving CS, and they revised the national exam (Matura) in Informatics. A new Subject examination catalogue was published in 2013 for the 2015 Matura examinations. This led to concern about the current textbook - does it support the change? Unfortunately, the answer is no. The CS textbook was first published in 1997, centered mainly on technology and its use (*Digital Literacy*). *Information Technology* and *CS as a discipline*, as defined in [19], was mostly left out. It was big and heavy, as it was supposed to cover all

A. Brodnik and F. Tort (Eds.): ISSEP 2016, LNCS 9973, pp. 167–178, 2016.
DOI: 10.1007/978-3-319-46747-4_14

four years. It was then re-published every two years, with minimal corrections. In 2015, teachers used the textbook, which was last revised in 2008, and contains some ridiculous content. Therefore, we saw the need for a new CS textbook.

This paper describes the development of the new textbook and presents the teachers first impressions after using it. First there is the Introduction section, followed by the section where the challenge is described in more detail. The alternative to heavy textbooks is presented and the progress in CS Education outlined. In the third section the new textbook is described, explaining its technology and content. The fourth section represents the results of a survey among the teachers. The paper finishes with the conclusion and intentions for the future.

2   The Challenge

The main challenge was a new educational tool. Why do we need it and what kind? Should it be a textbook or educational software? While virtual learning environments seem efficient and motivational enough, we decided to create an interactive textbook. The first reason for the decision is a necessary leap from the existing textbook, which was the Holy Grail for the majority of the teachers and emphasized computing literacy mostly. The second reason is the effort teachers are willing to put into their teaching. We noticed, that teachers prefer to use a textbook than educational software, because it usually takes less time to prepare for the teaching. For instance, project TOMO (https://www.projekt-tomo.si/, [17]) is a Slovene online educational software for learning programming, where teachers could create their own courses, they could copy or insert programming tasks and much more. In three years time they only created four courses, which are all empty, though. On the other hand using new virtual environments, however, run the risk of taking the *e-textbook* to become a collection of digital items, missing the main essence of what constitutes a textbook, as a *defined unit of content* with a clear message [18].

There are some online interactive textbooks: payable like **TeenCoder** for AP Computer Science A course in Java (http://www.compuscholar.com/teencoder/ teencoder_jv_series.php) and **Interactive Java** for elementary Java course (http://ijava.cs.umass.edu/index.html), free like **Computer Science Circles** for programming in Python (http://cscircles.cemc.uwaterloo.ca/) and even open source like **IMI Python** (http://imi.pmf.kg.ac.rs/imipython/) and **How to Think Like a Computer Scientist**, both for programming in Python (http:// interactivepython.org/runestone/static/thinkcspy/index.html).

In this section the background of the textbook is explained and an alternative is offered. Then the development of CS through important documents worldwide is described, and the situation in Slovenia is shown.

2.1   E-textbooks

Before we look at some characteristics modern e-textbooks have or should have, let us recall briefly what a textbook is. In [11] a textbook is defined as a *part of methodologically-didactical materials, and it cooperates with the teacher in the education process* and [20] emphasises that *The definition of a textbook depends on the nature of the school system. A textbook is one of the means that help the teacher and the student to achieve those goals.*

Besides the obvious additions and improvements to the *paper* textbook such as ease of access, lower weight and costs, speed of delivery, portability and ease of navigation [14], e-textbooks should provide content adapted to human to computer interaction with interactive elements, multimedia, instant feedback [9]. But this is not enough. An important aspect is missing. As the need for individual approach towards each student is becoming more and more accentuated, one of the crucial changes that is expected is that an e-textbook should allow for customization and personalization. Therefore, e-textbooks should be designed to be adaptable to the pedagogical situation and to the user, be it a learner or a teacher. It should be a given that an e-textbook allows for and enables uncomplicated customization and personalization.

What are the desired characteristics of a good e-textbook? According to [15, 16] they should be:

- **Accessible:** an e-textbook should be available online and there should be the possibility of transferring it to other locations.
- **Adaptable:** an e-textbook should be adaptable to the needs of individual teachers, learners and groups of learners.
- **Cost effective:** an e-textbook should increase the efficiency and productivity by cutting the time and money spent on the whole lifecycle of a textbook, including future revisions, adaptations.
- **Durable:** an e-textbook should be adaptable to the changes in technology without costly redesign and re-encoding.
- **Interoperable:** an e-textbook should have the option of being used in different learning environments and with different tools. Poor examples of this feature are some existing e-textbooks that require the use of a specific type of interactive whiteboards.
- **Reusable:** an e-textbook should have the option to use its parts in different contexts. For instance a teacher can use an applet in a frontal type lecture, and a student who makes a certain mistake while solving an exercise, is directed to that same applet. The exercises can be used as homework or as part of an exam. But the major point of being reusable is to use parts of different e-textbooks to produce a customized version of the e-textbook.

Perhaps the most important aspect of a future e-textbook is its adaptability. An e-textbook must be adaptable to the needs of individual teachers, learners and groups of learners. There is no real reason why the textbook used in class 7a should be the same as the one in class 7b, or even within the same class, why Joes textbook should be exactly the same as Janes. Some steps have already been taken in this direction. Several publishers offer the possibility of changing the order of the chapters; skipping and adding topics or even changing the contents.

Adaptability is the core idea and the key feature separating future e-textbooks from their paper (as well as from digitally enhanced) versions.

Moreover, it is the role of the teachers to exploit this adaptability. The teachers are the ones who must adapt the e-textbooks to an actual teaching situation and to a particular student. The authors create e-textbooks having a particular ideal situation in mind. The teachers, however, teach in the real world. Therefore, e-textbooks should be flexible. They should enable the teachers to change and recombine various parts from various sources. Unfortunately, in the majority of the existing e-textbooks, this is a mostly unrealized goal, although advantages in technology made this goal possible. However we should clarify the role of teachers in this adaptations. As textbooks reflect the academic standards, specific objectives, and ideologies commonly found in public curricula the role of the authors is to provide various models according to the foreseen pedagogical situation and thus provide teachers choices [22]. Techers will pick the most appropriate version of the e-textbook and personalize/adapt it further.

2.2   Curricula

Although Computer Science is a relatively young science, its evolution was quite turbulent. From the theoretical Turing machine, through electronic general-purpose computer and programming languages, to personal computer and the goal to teach people how to use this new technology. Schools started to introduce Computer Science (CS) to students. Unfortunately, with the creation of new technology, it was necessary to learn how to use it. In most schools the CS course was slowly replaced by an ICT course, and in some cases that even escalated to a course in the use of office tools. Various researchers opposed the new trend and in the last 10 years CS concepts aim to find a way back to the curricula.

One of the first important documents about CS Education is probably IFIP's ICT curriculum in secondary education from 1994 (updated in 2000). While it basically introduces ICT literacy and basic skills into the school, it mentions creating and supporting of ICT, but is not intended for general education, but for professional education [4]. For the next decade schools mainly thought ICT literacy.

The first attempt to break this period was A Model Curriculum for K-12 Computer Science in 2003. Its main goal was to introduce the principles and methodologies of CS to all students. They specify the distinctions between CS and information technology, and recommends structure for K-12 curriculum [7]. For 5 more years, nothing much has changed until the breakthrough, which was almost simultaneous in 3 different continents.

In New Zealand prof. Tim Bell with his colleagues made a collection of free learning activities that teach CS through engaging games and puzzles that use cards, string, crayons and lots of running around, named *CS Unplugged* [3].

In the USA they published a national report *Running on Empty: The Failure to Teach K-12 Computer Science in Digital Age*. In the report they present findings about the poor situation of CS in US schools and list recommendations for improving CS education [23]. Soon after the report, CSTA published renewed

K-12 Computer Science Standards, where they emphasize the role of CS as a core discipline and restructured K-12 curriculum [8].

In the UK they decided on a more aggressive approach with the report called *Shut down or restart? The way forward for computing in UK schools*. The main findings of this comprehensive report include the unsatisfactory CS education, the role of CS as an academic discipline and poor qualifications. They thoroughly describe issues and recommendations, and suggest terminological reform ICT is divided to CS, Information Technology and digital literacy [19]. In the following years, a new compulsory subject Computing was introduced to all schools. With the help of Computing At School (CAS) organization, new materials for teachers and students were made, along with a new curriculum [5,6].

The latest project of renewing the CS begun in the end of 2015 by a group of various participants, from organizations ACM, CSTA and Code.org, to schools and technology companies. Their goal is to create a general framework for K12 CS, which would identify core concepts and practices. It is not meant to be a curriculum or standards document, but instead to provide the guidelines for designing a new curriculum, assessments or teacher preparation programs. In the beginning of 2016 they published *A Framework for K-12 CS Education*, a draft of framework, consisting of five core concepts and seven practices. The draft was published for open reviewing by anyone who wanted to support the development of the framework and help to improve it. The final version will be finished and published in the summer of 2016 [12].

In Slovenia we also came to a conclusion that the situation in schools is alarming and change needs to be done. The Government and the National Education Institute agreed to create a new (elective) Computer Science subject in grades 3, 4 and 5. The new subject covers topics from CS Unplugged. We still have an elective CS-like subjects in grades 6, 7 and 8, which remain the same, a mandatory CS subject in grade 9 (gymnasium) and an elective CS Subject in grades 10, 11 and 12. Overall we have one mandatory year of CS and nine optional years of CS in K-12 school system.

While the curriculum itself could not be revised, it is fortunately quite open. Its openness actually goes in two directions a positive and a negative one. The positive side of the curriculum is of course the fact, that teachers can include all the desired topics into the subject, with no set limits. They can be creative, they can introduce the students to physical computing, they can do all sort of things. Our curriculum is made for the whole secondary school (4 years) with no distinction between the years. A teacher can therefore decide what he/she will teach and when he/she will teach it. Unfortunately, in practice that means, that a teacher teaches "easier topics", such as the use of programs and/or computer in the mandatory subject in the first year, and leaves the important topics, such as algorithms, programming and networks for later (elective) courses. This means that only a small percentage of secondary school students get familiar with the proper content. Why there is such a small percentage of these students, is a story for another article. The first step toward a better subject was made with the publication of the Computer Science examination catalogue for Matura, which included ACM's knowledge areas. However, teachers were struggling with the new content, because the old textbook did not cover all the topics. The next logical step was a new textbook, an interactive online textbook.

3   New E-textbook

We immediately decided that the form of the new textbook will be interactive, it will be online and free - quite the opposite from the old traditional one. We had the support of the ministry as well as the National Education Institute.

We faced the first challenge, where we had to decide which topics will be included in the textbook. We quickly saw, that all of the areas cannot be included in just one textbook, because that would be too much. Therefore we decided, that the four most important topics: programming and algorithms, systems, networks and distributed systems and informatics and society, need to be in the first part of the textbook.

We wanted to make a clear and concise textbook, with less text, but with comprehensive explanations. Therefore, the second challenge was to balance the explanations and the limited space in the textbook, and to include just the right amount of interactive elements [2].

The new interactive online textbook is meant for students, but can also be used by teachers. In the following sections we will explain the technology behind the textbook and its limitations, the importance of learning objectives from the curriculum and the grouping of areas.

3.1   Technology

The interactive textbook was created with a tool called *exeCute* (http://execute. fnm.uni-mb.si/). The tool is an adapted version of *eXeLearning tool* (http:// exe-learning.org/), which originates in New Zealand. It is a freely available Open Source authoring application to assist teachers and academics in the publishing of web content. The tool was upgraded by a Slovene team into an *exeCute tool*. They used their own XML scheme that describes the content in a neutral XML format called E-learning object XML.

As seen in Fig. 1, it supports different kinds of interactive elements which can be made with JavaScript, some are even predefined:

– low-level interactivity elements, such as images, video, sound, animation, simulation (multimedia components);
– medium-level interactivity elements, such as various tests (true/false, multiple-choice questions, gap-fill);
– high-level interactivity elements, such as applets and educational games.

The tool itself did not support built-in code interpreter, therefore we needed to add it ourselves. With the help of *CodeMirror* (https://codemirror.net/) and *Skulpt* (http://www.skulpt.org/) libraries, we added an interactive and a non-interactive interpreter directly into the e-textbook, so pupils can code alongside

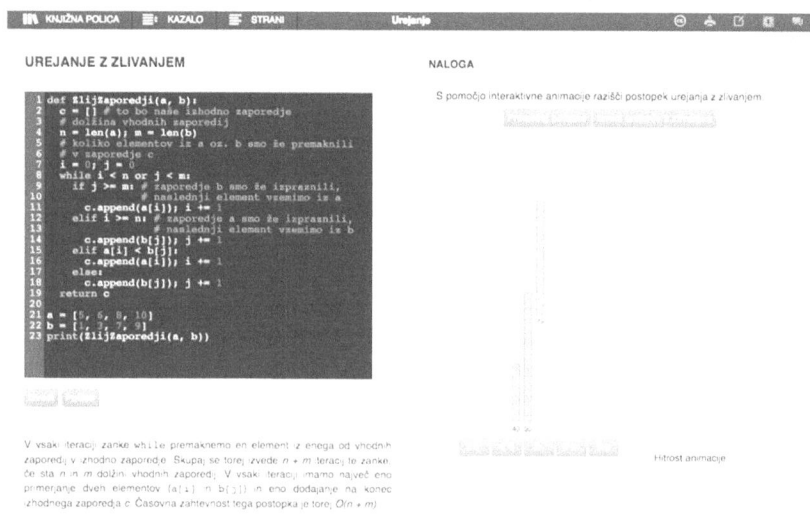

Fig. 1. Explanation of MergeSort, code and example of simulation.

the content or the exercise. *CodeMirror* serves as the base for the editor (line numbers, syntax highlighting, etc.), while *Skulpt* interpret Python code into the JavaScript code, execute it and return the result.

ExeCute has an index structured content covering each learning set that contains learning units - which are the base of the textbook. An e-learning unit consists of the title, the introduction (motivation and presentation of the content), the body (main content and interactive elements), the conclusion (summary of what the student learned and a set of exercises), and the sources [21].

A lot of interactive textbooks had already been made with the *ExeCute* tool by various authors. In Table 1 we describe the positive and the negative properties of the tool, based on our experience.

Moreover, the ExeCute technology supports modularization of learning objects and their grouping. Since it is XML based it is easy extendable by introducing new attributes. Due to lack of time we were not able to fully employ them, but the plan is to use additional attributes to support the personalization of the learning path and consequently make our e-textbook fully adaptable.

3.2   Grouping of Areas and Learning Outcomes

In Slovenia we have a CS curriculum for secondary school (gymnasium) from 1998, which was revised in 2008. There are merely 15 pages, although it covers all four years (from grade 9 to grade 12 in K12 system). The most interesting part is the distribution of topics with their 100 learning objectives. As mentioned in the beginning, the teacher can choose which topics he/she will teach in the one mandatory year, and which ones in the later optional years. Or in other words,

Table 1. Properties of the *ExeCute* tool.

PROS	CONS
- simple graphical user interface	- run only on Windows OS
- easy to use	- poor documentation
- online support of the authors	- slow, if you have a large e-learning unit
- nice design	- you can open only one e-learning unit per session
- predefined medium-level interactivity elements	- if you want to close and stop the program, you need to kill its process and Mozilla Firefox processes by hand
	- the prosess of inserting JavaScript applets is not straight-forward
	- complicated process of exporting the e-learning unit with all the included functionalities

which learning objectives should be presented to all students. The question is, how could one choose the most important topics and the most important learning objectives out of 100? In practice, the easier topics always win. Of course, the formal reason for choosing easier topics is much smarter - with the easier topics the teachers cover 54 % of all CS content, in one year. Sadly, 67 % of the easier topics are about the use of different programs.

Although we agree, that *digital literacy* is also important part of CS, this should not be the focus of the CS course. Computing is focused on three components: Digital literacy, ICT and Computer Science as a discipline. The last one is the least represented or even left out, so we will focus mainly on this component. A very good set of the knowledge areas in CS is presented by ACM and IEEE Computer Society [1]. After analysing other curricula and standards documents, authors of the e-textbook, including several university professors and secondary school CS teachers, incorporated programming and algorithms, systems, networks and distributed systems, and informatics and society into the new e-textbook. Throughout the e-textbook there is an emphasis on computational thinking.

The new e-textbook was published online (http://lusy.fri.uni-lj.si/ucbenik/) in the middle of 2015 and presented to all CS teachers in Slovenia. We were really satisfied with the result, especially when we found out that the content of the new e-textbook corresponds to the concepts of the new guidelines of A Framework for K-12 CS Education [12].

4   First Impressions

The new e-textbook was made and it was presented to most of Slovene CS teachers. We wanted to know, if the teachers use the e-textbook, how they use it, and what they think about it. We created a short online survey and received 61 replies. The complete questionnaire with analysis is available online (http://lusy.fri.uni-lj.si/ucbenik/survey2016).

The survey was divided into four sections - **Basic information**, **Content**, **Technology**, **Impact on learning process**, and **Comparison with the old textbook**.

Out of 61 replies, there were 35 (57,4 %) male and 26 (42,6 %) female teachers. 7 (11,5 %) of them were less than 30 years old, 14 (23 %) between 30 and 40, 20 (32,8 %) between 40 and 50, and 20 (32,8 %) more than 50 years old. 11 (18 %) teachers have just started teaching and have less than 5 years of experience, 6 (9,8 %) teachers have between 5 and 10 years of experience, 12 (19,7 %) teachers have between 10 and 15 years of experience, 14 (23 %) teachers have between 15 and 20 years of experience, and 18 (29,5 %) teachers have more than 20 years of experience. 5 (8,2 %) of them were not familiar with the e-textbook before, all others (91,8 %) were familiar with it.

We can see that the majority of replies was made from senior teachers, over 40 years old (65,6 %) and with more than 15 years of experience (52,5 %).

As expected, more than half - 34 (55,7 %) teachers teach at gymnasium, 18 (29,5 %) teach in primary school, 7 (11,5 %) teach in secondary professional schools and 2 (3,3 %) teach elsewhere.

The Content and Technology parts included three questions each; one question with a scale and two open questions. In the **Content** part, the question with a scale read *The content is explained well* and teachers could choose numbers between 1 (Strongly disagree) and 4 (Strongly agree). We chose the short Liker scale (4-point) in order to eliminate the *I do not know* answer. We wanted the teachers to decide whether they agreed or not. Most of them, 95,1 %, agreed that the content was explained well.

We asked the teachers which topics they considered important and felt they needed to be added. Most teachers missed topics about digital literacy (22 %), a lot of them suggested all the missing topics from the curriculum (20 %), only one teacher suggested the topic about physical computing and only one teacher suggested the topic about mobile applications.

We expected them to miss the digital literacy part from the curriculum, but did not expect so few different suggestions.

In the **Technology** part, the question with a scale read *E-textbook is technologically well made and has proper interactive elements*, where we chose the short Likert scale again. The majority agreed with the statement (93,5 %). The interactive element that the teachers liked most was animation (39 %), the second one was the built-in Python interpreter (28 %) and the third element was tasks with feedback (25 %). We got a few interesting answers to the open question if they would change anything from the aspect of technology. Two teachers suggested better support for mobile and tablet browsers, one teacher suggested a built-in search field, one teacher suggested support for adding notes. However, four teachers suggested more interactive elements, as the current version was too serious and boring, and one teacher suggested a printed version of the e-textbook, so that the students would not look into the display for too long.

In the **Impact on learning process** part, we first asked the teachers if they use the e-textbook in class, and then, based on the answer, provided sub questions.

Half of them, 31 (50,8 %), use the e-textbook, and 30 (49,2 %) of them do not use it. From those who use it in class the most common two descriptions of usage were exercises and revision. We grouped similar answers into five groups, as seen in Fig. 2. 19 % of the teachers use the e-textbook for explaining the concepts with interactive elements, 38 % of them use it for exercises and for revision of the content knowledge, 24 % of them improve their own learning plans and slides, 9,5 % use it for flipped learning, and 9,5 % for additional explanation. From those teachers, who don't use the e-textbook in class, 7 (23,3 %) of them said they didn't know about it yet but plan to use it in the future.

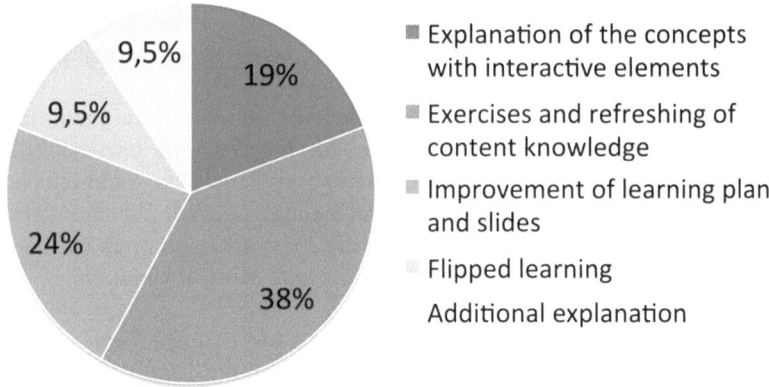

Fig. 2. Distribution of usage of the e-textbook.

The last part was a **Comparison with an old textbook**. 18 (30 %) teachers did not use any textbooks, they prefer to find information on the internet, 40 (66 %) teachers used the previously mentioned textbook and 3 (4 %) teachers used other textbooks. The majority of teachers - 51 (84 %) find the new e-textbook more useful, and the content better explained, although in the open question section they emphasize the lack of topics from the curriculum.

We can conclude, that the teachers liked the new e-textbook, both the content and the technology. However, they need to follow the curriculum and explain all the topics, which means the e-textbook is not yet complete.

5   Conclusion and the Future Steps

As teachers are mostly the ones who decide which textbooks are to be used in their classrooms, they are the ones who influence textbook development [13]. On the other hand, this also means that a careful choice of the approach used in a textbook towards the topics prescribed by curriculum can realize some shifts in the way computer science is taught in Slovene schools.

According to the results of the preliminary study of the teachers' opinion described in this paper the authors of a new Slovene Computer Science e-textbook are on the right track. The topics covered are in accordance with the latest developments of curricula of computer science, the teachers prefer the technological solutions used with plenty of interactivity, especially in programming and algorithms.

However, the present version should be mostly seen as the first step towards a really modern e-textbook for Computer Science. The aspect most needed, and unfortunately missing in the current version, is adaptability. As argued in the introduction, e-textbooks should be designed to be adaptable to the pedagogical situation. There are several reasons why this goal has not yet been achieved in the current version of the Slovene Computer Science e-textbook. The most important is the fact that the Computer Science e-textbook was meant to be included in the Slovene portal of e-textbooks [10]. This brought some requirements and limitations of the layout as well as of the technological solutions that could be used. Due to some constraints the textbook described here was not actually included in the portal, so in the future versions some more advanced approaches in technological preparations will be used. Fortunately, as content and technological solutions currently used are based on HTML, the adaptations are possible. Besides adding adaptability we are planning to add support of other programming languages, not just Python.

References

1. ACM/IEEE-CS Joint Interim Review Task Force: Computer science curricula 2013: Curriculum guidelines for undergraduate degree programs in computer science (2013). https://www.acm.org/education/CS2013-final-report.pdf. Accessed 4 May 2016
2. Anželj, G., Brank, J., Brodnik, A., Bulič, P., Ciglarič, M., Djukič, M., Fuerst, L., Kikelj, M., Krapež, A., Medvešek, H., Mori, N., Pančur, M., Sterle, P.: Computer science and informatics 1. e-textbook for informatics in gymnasium (2015). http://lusy.fri.uni-lj.si/ucbenik/. Accessed 4 May 2016
3. Bell, T., Alexander, J., Freeman, I., Grimley, M.: Computer science unplugged: school students doing real computing without computers. J. Appl. Comput. Inf. Technol. **13**(1), 20–29 (2009)
4. Bosler, U., Gumbo, S., Taylor, H., Wati Abas, Z., Duchteau, C., Morel, R., Waker, P.: Information and communication technology in secondary education. A Curriculum for Schools. UNESCO (1994)
5. Computing At School: Computing in the national curriculum. A guide for primary teachers, CAS (2013)
6. Computing At School: Computing in the national curriculum. A guide for secondary teachers, CAS (2014)
7. CSTA: a model curriculum for k-12 computer science (2003). http://www.acm.org/education/curricula-recommendations. Accessed 4 May 2016

8. CSTA: Csta k-12 computer science standards (2011). http://csta.acm.org/Curriculum/sub/K12Standards.html. Accessed 4 May 2016
9. Daniel, D., Woody, W.D.: E-textbooks at what cost? performance and use of electronic v. print texts. Comput. Educ. **62**, 18–23 (2013)
10. iUčbeniki: iučbeniki - spletno mesto interaktivnih učbenikov(itextbooks - portal of interactive textbooks) (2016). http://eucbeniki.sio.si/index.html. Accessed 4 May 2016
11. Jurman, B.: Kako narediti dober učbenik na podlagi antopološke vzgoje. Jutro, Ljubljana (1999)
12. K12CS.org: a framework for k-12 computer science education (2016). https://k12csdotorg.files.wordpress.com/2016/03/k-12-cs-framework-draft-march-18-2016.pdf. Accessed 4 May 2016
13. Knecht, P., Najvarov, V.: How do students rate textbooks? A review of research and ongoing challenges for textbook research and textbook production. J. Educ. Media Memory Soc. **2**(1), 1–16 (2010)
14. Lai, J., Chang, C.: User attitudes toward dedicated e-book readers for reading: the effects of convenience, compatibility and media richness. Online Inf. Rev. **35**(4), 558–580 (2011)
15. Lokar, M.: The future of e-textbooks. Int. J. Technol. Math. Educ. **22**(3), 101–106 (2015). Burnham: Research Information Ltd
16. Lokar, M.: E-textbook of the future. In: Proceedings of the Time 2014, R&E-SOURCE (2016). http://journal.ph-noe.ac.at/index.php/resource/article/view/135. Accessed 4 May 2016
17. Lokar, M., Pretnar, M.: A low overhead automated service for teaching programming. In: Proceedings of the 15th Koli Calling Conference on Computing Education Research, pp. 132–136. ACM, Koli Calling 15 (2015)
18. MindCET: The future of digital textbooks (2012). http://www.mindcet.org/wp-content/uploads/2012/10/Digital-Textbooks.-A-literature-review1.pdf. Accessed 24 July 2016
19. The Royal Society: Shut down or restart? The way forward for computing in uk schools (2012). http://royalsociety.org/education/policy/computing-in-schools/report. Accessed 4 May 2016
20. Turk Škraba, M.: Učbenik kot sredstvo za kakovostno učenje in poučevanje družboslovja. Ljubljana: diploma thesis (2005)
21. Čuk, A., Drakulič, D., Flogie, A., Jelen, S., Kaučič, B., Lipovec, A., Milekšič, V., Mohorčič, G., Novoselec, P., Pesek, I., Prnaver, K., Regvat, J., Repolusk, S., Senekovič, J., Šenveter, S., Vrtačnik, M., Zmazek, B., Zmazek, B., Zmazek, E., Wassermann, A.: Slovenian i-textbooks. The National Education Institute Slovenia, Ljubljana (2014)
22. Väljataga, T., Fiedler, S.H.D.: Going digital: literature review on E-textbooks. In: Zaphiris, P., Ioannou, A. (eds.) LCT 2014, Part I. LNCS, vol. 8523, pp. 138–148. Springer, Heidelberg (2014)
23. Wilson, C., Sudol, L.A., Stephenson, C., Stehlik, M.: Running on empty: the failure to teach k-12 computer science in digital age (2010). http://www.acm.org/runningonempty/. Accessed 4 May 2016

Computer Science in the Eyes of Its Teachers in French-Speaking Switzerland

Gabriel Parriaux[✉] and Jean-Philippe Pellet[✉]

University of Teacher Education, Lausanne, Switzerland
{gabriel.parriaux,jean-philippe.pellet}@hepl.ch

Abstract. This paper discusses the situation of high-school-level Computer Science education (CSE) in the French-speaking part of Switzerland through the eyes of Computer Science teachers. After presenting the peculiarities of the educational system in a federal state like Switzerland and its impact on CSE, we try to answer several questions about CS teachers, their profile, and their representations of the field. Recognizing that the primary field of study of most current CS teachers was not CS, we question their representations of CS in search of potential differences between specialists and non-specialists. On this basis, we analyze the distance between CS as it is taught in French-speaking Swiss high schools and CS as its teachers think it should ideally be taught. Finally, we present the important need for continuing education of CS teachers and the fact that, according to them, it should include both technical and didactic aspects.

Keywords: Computer science · Computer science education · Computer science teachers · Swiss high schools · Representations of the field · Continuing education

1 Introduction

This paper is concerned with several issues linked to teaching Computer Science education (CSE) in French-speaking Swiss high schools, presented according to the following structure. Section 2 presents the characteristics of the Swiss educational system, its impacts on the organization of CSE in high schools and the situation of CS teachers. Section 3 outlines our research questions and our methodology to collect data. We then present and discuss our results in Sect. 4 and sum them up in the conclusion as Sect. 5.

2 Historical Elements and Context

Switzerland is a federal state composed of 26 cantons and half-cantons. Since their origins, Swiss people have considered very important to give cantons a lot of independence from the federal state. It is apparent in a lot of dimensions: political, economic, educational, to name a few. This organization has a lot of

© Springer International Publishing AG 2016
A. Brodnik and F. Tort (Eds.): ISSEP 2016, LNCS 9973, pp. 179–190, 2016.
DOI: 10.1007/978-3-319-46747-4_15

positive aspects, letting political decisions be taken by people who are close to the field, but also less positive ones, leading to a greater complexity.

This also holds for education. Education is mostly managed at a cantonal level, which means that Switzerland has nearly 26 different educational systems with 26 education ministers. Some processes and instances do exist to try and coordinate decisions and systems between cantons, but nevertheless education politics remains complex to understand.

The Case of High Schools. Even if high schools depend from the cantons, students obtain a so-called "federal maturity" when they graduate from them— "federal" meaning that it is valid in the whole country.

The country-wide recognition of high-school diplomas is regulated by a federal document (hereafter referred to as "RRM"[1]). Cantons must conform to the rules listed in RRM in order for their diplomas to be validated by the state [4]. RRM establishes globally the fields that must be taught in high schools along with the rules for certification. In more details, it distinguishes four main teaching domains (languages, mathematics and sciences, humanities, and arts) and three lists of disciplines: (*a*) fundamental fields, which must be taught to all students; (*b*) so-called "specific options," which can be viewed as the high-school version of college majors; and (*c*) complementary options. Students have to choose a single specific option and a single complementary option; therefore, each of them only concerns a (possibly small) subset of students. RRM doesn't dictate the number of teaching periods assigned to each field, but only gives an indicative proportion of each of the four domains. It also doesn't describe the contents of the fields. Cantons have the liberty to propose canton-specific disciplines in addition to the RRM-mandated ones.

In this context, RRM is the most important document that exists. The version of RRM valid today was written in 1994, with some adjustments made in 2007.

With RRM having established the fields of teaching and learning, there is a second document (hereafter referred to as "PECMAT"[2]) established by the Swiss Conference of Cantonal Ministers of Education ("CDIP"[3]). It describes a short "curriculum framework" for each discipline mentioned in RRM. It is not legally binding, but makes recommendations to the cantons [2]. PECMAT dates to 1995 and a complementary part was written in 2008 to reflect the changes introduced in RRM in 2007.

Based on PECMAT, the cantons each establish their own operational curricula, which serve as reference for teachers. The process ends here with 26 cantonal curricula for each field (for instance, [5,7]).

---

[1] *Règlement de reconnaissance des maturités* or *Anerkennung von gymnasialen Maturitätsausweisen.*

[2] *Plan d'études cadre pour les écoles de maturité* or *Rahmenlehrplan für die Maturitätsschulen.*

[3] *Conférence suisse des directeurs cantonaux de l'instruction publique* or *Schweizerische Konferenz der kantonalen Erziehungsdirektoren.*

Computer Science in High Schools. In the 1994 version of RRM, CS didn't exist as a field, but was only mentioned as a collection of transdisciplinary topics. In the period from 1994 to 2007, considering the lack of CS or related field in the federal rules, some cantons decided to make use of their freedom to introduce CS as a cantonal field.

There are no studies about the motivations of the cantons to introduce CS as a cantonal field at that time, so uncertainty remains as to how this process precisely occurred. Certain is that it was made independently of any federal recommendations, so each canton decided on its own on the contents to be taught. Without aiming at providing a detailed look at those cantonal curricula (which would be outside our current scope), a quick look at them reveals that the contents of this field called "Computer Science" (*informatique* in French) is closer to teaching and learning the use of traditional software tools (word processing, spreadsheets, etc.) than to the academic discipline as we identify it today. It seems that the preoccupation of education ministers at that time was to make sure that students were able to produce proper presentations, written texts and graphs for their school work. If it were done today, we would certainly question the relevance of the name of "Computer Science".

In the 2007 addendum to RRM, CS was introduced at a federal level as a new discipline in the list of complementary options. For the first time, the opportunity was given to students to study CS as a scientific field. An addendum was written to PECMAT to propose a description of the contents of this new course and, in a typical process for Switzerland, each canton wrote its own operational curriculum. A quick look at the PECMAT addendum or at the cantonal curricula derived from it shows that the mentioned themes are closer to CS as a scientific field and not so much related to the use of software tools.

The addition of CS as a complementary option was considered a major improvement by people concerned by the state of CSE in the country. But owing to the nature of complementary options, only a few students actually got to study CS that way and the concrete impact of this new course was thus limited.

In reaction to the introduction of CS as a complementary option in RRM in 2007, a few cantons decided to suppress the CS cantonal field they had introduced before, but the majority of them kept both. Today, the resulting situation can be characterized this way: very diverse depending on the canton, with mostly two kinds of CS courses side by side in the curricula: one cantonal with an emphasis on the use of software tools (referred to later as "cantonal CS"), and one federal with a scientific orientation (referred to later as "complementary-option CS")—both of them named "Computer Science".

In 2013, CDIP gave mandate to one of its subgroups to write a report about the introduction of CS in high schools as a fundamental field for all students. In this mandate, CDIP clearly states that the presence of CS in high schools must be strengthened in regards to its importance in society nowadays [3]. As we write this article, work towards the final report is reportedly in progress. If that report recommends the introduction of CS as a fundamental field, political decisions will need to be made in order to adapt the structure of the domains and curricula in high schools, as well as the official documents (RRM and PEC-MAT). Understandably, said mandate generated high expectations among CSE professionals, who see a true opportunity for the introduction of CS for all students in Swiss high schools in the near future. The impacts of such a decision could be very important, in particular for CS teachers.

Situation of Teachers. In the 80s, computers were introduced in Swiss schools before any CS curriculum existed. Teachers who graduated in CS didn't exist either. CS curricula were not so widespread in universities and as there was no CS in schools, there was no reason for a CS specialist to work as a teacher. Often, mathematics teachers or physics teachers (sometimes teachers of other fields) got in charge of managing the school computers because they were the only ones who had ever seen computers during their college studies. Quite naturally, when some cantons later introduced CS curricula in their schools, those same teachers started teaching it. Even if it is a mandatory rule that high-school teachers must hold a Master's-level degree in their field of teaching [4], at the time, a margin of tolerance existed, supposedly due to the fact that CS was canton specific.

Things gradually changed and starting around 2000, more students holding a Master's degree in CS have been seen entering teacher-education programs and becoming CS teachers in high schools.

When CS debuted as a complementary option in RRM in 2007, there was an important need for CS teachers with an academic background in CS. An ad hoc continuing-education program in CS was proposed to non-specialist teachers who were already in charge of the cantonal CS course. Between 40 and 50 teachers graduated from that program.

3   Research Questions and Methodology

In short, the situation of CS in Swiss high schools is a bit confusing: the federal course as a complementary option coexist with the cantonal course, both being named "Computer Science", but with different contents. Some curricula focus on the use of software tools while others are closer to academic CS. CS teachers have different profiles, some of them being CS graduates, some others being primarily specialists of other fields. In addition, each canton has its own organization and curricula.

As we might be on the cusp of major change with the potential introduction of CS for all at a federal level, there is a need to clear up the confusion and get a better understanding of the current situation.

We decided to focus our efforts on the following research questions:

1. *What is the proportion of CS teachers who primarily graduated in CS?*
2. *Do teachers with different backgrounds view CS fundamentally differently?*
3. *What are the differences between ideal CS teaching, complementary-option CS, and cantonal CS in the eyes of the teachers?*
4. *What topics do CS teachers need in continuing education?*

Our method of investigation is based on a survey addressed to CS teachers. As our institution is involved in the education of teachers for the French-speaking part of Switzerland, we focused on that part of the country. The survey was sent to teachers through one of the most important professional associations of CS teachers in Switzerland, the Swiss Society for Computer Science in Education (SSIE[4]).

The survey was composed of four parts: 1. teachers' profile (academic and pedagogical studies, current teaching); 2. needs for training; 3. representations of CS and CSE; 4. opinion on a potential CS course for all students.

In order to better design our survey, we ran a preliminary version of it during three personal interviews with three CS teachers who had different profiles and backgrounds. We then proceeded to some adjustments to better fit our goals.

4 Results and Discussion

The total number of respondents was $N = 37$. The population size (i.e., the exact number of CS teachers in French-speaking Swiss high schools) is not known to us as we could not readily obtain such information from the cantons, but we estimate it to be between 150 and 200.

Like for most surveys based on voluntary participation, the sample formed by the respondents may be biased in several ways. We expect teachers with an interest in the development of CS teaching to be more likely to participate. In particular, we noted a large representation of a special subpopulation: teachers who participated in the CS continuing-education program mentioned at the end of Sect. 2, offered when complementary-option CS was introduced. We also expect teachers in need of continuing education to be more likely to want to give their input. Finally, we had no way of ensuring that every member of the population would effectively be notified of the survey.

Question 1. Figure 1 shows the initial fields of study of the respondents (as multiple answers were possible, the numbers add up to more than 37). Although most of them (31, about 84 %) primarily studied at least one STEM[5] field, only a minority (15, about 41 %) studied CS.

Older teachers are less likely to have primarily studied CS—as mentioned before, an obvious reason is that CS curricula were not as widespread as they have gradually become now. Since the late 90 s especially, a growing number of CS curricula have been proposed, a lot of them by the newly appointed universities of applied sciences[6]. We actually found out that the proportion of CS graduates was substantially larger for teachers who graduated after 2000: 8 out of 11 (73 %) vs. 7 out of 25 (28 %) for those who graduated before 2000.

---

[4] *Société suisse pour l'informatique dans l'enseignement* or *Schweizerischer Verein für Informatik in der Ausbildung.*

[5] Science, technology, engineering, and mathematics.

[6] *Hautes écoles spécialisées* or *Fachhochschulen.*

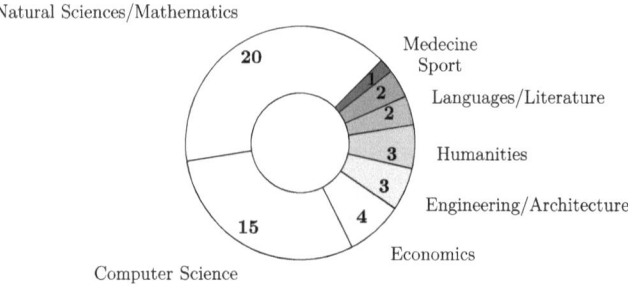

Fig. 1. The fields of initial studies of the respondents.

Question 2. Do teachers with different backgrounds view CS fundamentally differently? We asked respondents to indicate, for each of these items, whether they completely disagree, somewhat disagree, somewhat agree, or completely agree with it.

"Absolutely spoken, outside schools, computer science...

1. is mainly applied mathematics" (hereafter referred to as the AppliedMath subquestion)
2. doesn't really have stable components and changes all the time" (NotStable)
3. changes rapidly, but rests on stable notions that do not change a lot" (StableNotions)
4. has theoretical foundations" (HasTheory)
5. focuses mostly on abilities to use software tools" (Tools)
6. mainly represents know-how rather than concepts and notions" (KnowHow)
7. is the major science of the 21st century" (MajorScience)

Looking qualitatively at the respondents' education profiles, we categorized them into three groups: (G_1) those whose primary education was CS $(N_{G_1} = 15)$; (G_2) those whose primary education was not CS, but who had complementary or continuing CS-related education $(N_{G_2} = 17)$; (G_3) those who had no CS-related education other than being self-taught $(N_{G_3} = 5; N_{G_1} + N_{G_2} + N_{G_3} = N = 37)$. Comparative results on each subquestion, for each of the three groups and for all respondents together, are shown in Fig. 2.

These results show the following. 1. CS is considered by more than 80 % to be more than just applied mathematics. 2. Less than 5 % think that CS doesn't have stable components. 3. All but one respondent somewhat or completely agreed that CS rests on stable notions. 4. Less than 5 % disagreed that CS has theoretical foundations. 5. About 20 % are of the opinion that CS is mainly about how to use software tools. 6. Most (more than 80 %) disagree that CS mainly represents know-how. 7. More than 75 % somewhat or completely agree that CS is the major science of this century.

Although small group differences can be observed, Kruskal–Wallis H tests [6] conducted for each subquestion revealed that only subquestions HasTheory $(H(2) = 8.37, p = 0.015)$ and KnowHow $(H(2) = 6.71, p = 0.035)$ exhibited

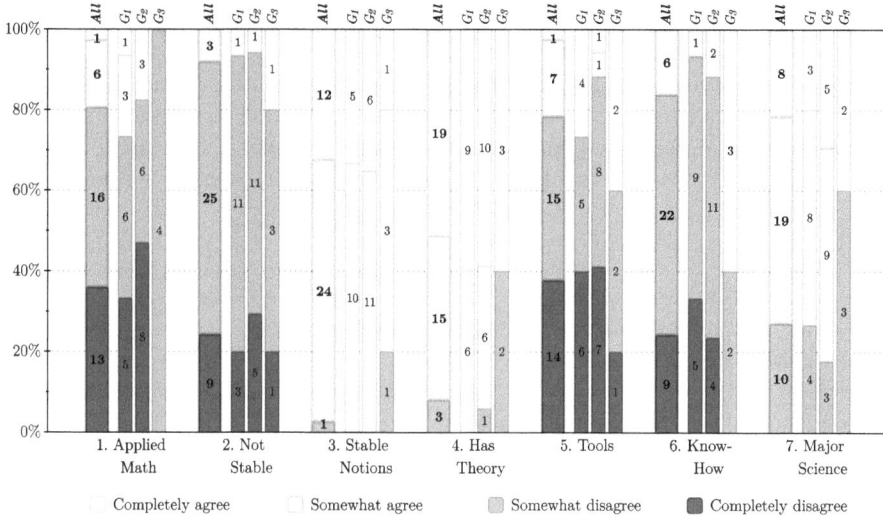

Fig. 2. The respondents' declared agreement on the nature of CS on 7 axes. Data is shown for the whole sample and for the three discussed subgroups.

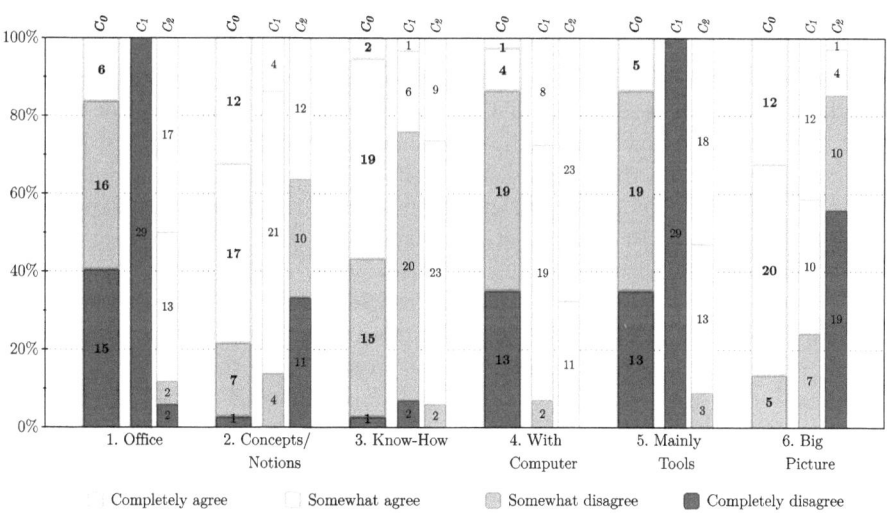

Fig. 3. The respondents' view on what CS teaching should be ideally (C_0) vs. what it is for two course types currently given in high schools (C_1 and C_2).

statistically significant differences between our three groups. In the former case, the self-taught group was significantly less likely to agree that CS has its own theoretical side; in the latter, they were significantly more likely to agree that CS was rather about know-how than concepts and notions. However, the small sample size of that group makes these results subject to caution.

Question 3. Teachers have an opinion of what the ideal format of CS teaching should be. We wanted to compare this ideal representation with the two course types that are currently given, namely, the complementary-option CS course (hereafter referred to as C_1) and the cantonal CS course (C_2). We thus asked respondents to indicate, for each of these items, whether they completely disagree, somewhat disagree, somewhat agree, or completely agree with it—once for C_1, once for C_2.

"CS teaching in the context of this course...

1. is mainly about learning how to use office software" (Office)
2. builds on concepts and notions" (ConceptsNotions)
3. consists mostly of know-how" (KnowHow)
4. is given with a computer rather than with paper/pencil" (WithComputer)
5. mainly has the purpose of teaching tools useful for the students' work" (MainlyTools)
6. gives a representative overview of what the academic discipline is" (BigPicture)

We then asked a similar question: "Ideally, CS teaching in high schools..." with the same six subquestions as mentioned above, in a "should" form (i.e., item 1. becomes "should mainly be about learning how to use office software," 2. becomes "should build [...]", etc.). We refer to this hypothetical ideal course as C_0 and compare the responses to those given for C_1 and C_2.

The results are shown in Fig. 3. Looking at the C_0 bars, we can say that for about 80 % of respondents, a CS course in high school does not concern itself with teaching how to use office or other software tools. It should build on concepts and notions that do not systematically require the involvement of a computer, and provide a representative overview of the discipline. Respondents are split on the KnowHow subquestion, with about 57 % only agreeing that an ideal CS course should mainly consist of know-how.

In an effort to better visualize the differences between the ideal case and the two course types currently given, we performed a principal component analysis (PCA, see e.g. [1]) of these answers. The scree plot of the PCA is shown in Fig. 5, and the answers, divided into three groups, are shown along the first two principal components in the scatterplot in Fig. 4. The projection of the 6 initial dimensions have been overlaid on the scatterplot in order to better understand the nature of the principal components.

The scree plot shows the large importance of the first component, while the first two explain almost 80 % of the variance. This gives us confidence in the faithfulness of the scatterplot representation, on which the three groups of points are quite clearly separated. The C_1 and C_2 groups are even linearly separable. The former has negative values along the first component, corresponding to a teaching oriented towards concepts and notions and an overview of the discipline; the latter is strongly oriented towards office and other software tools and know-how. Both have positive values along the second component, which translates to these courses being very often given in computer rooms, in interaction with hardware.

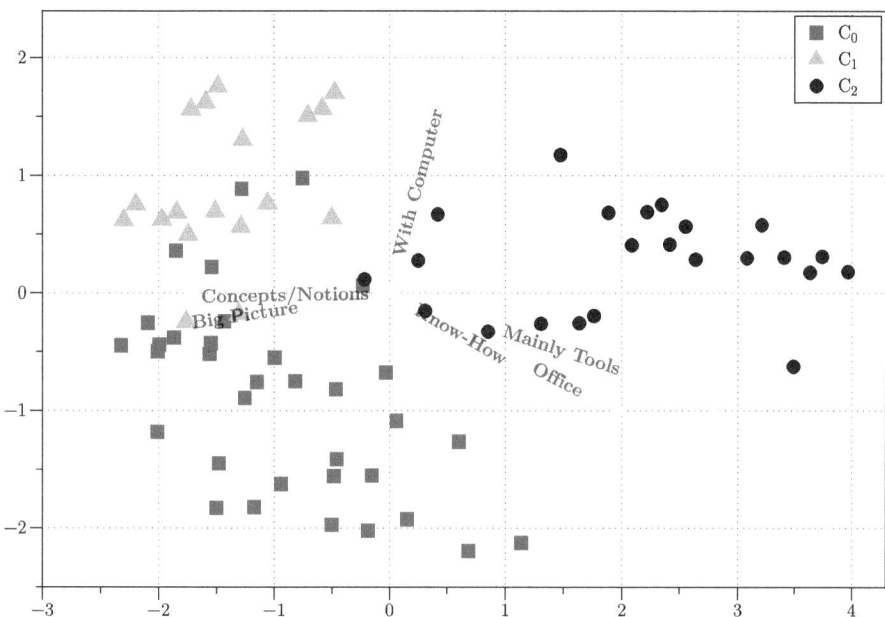

Fig. 4. Scatterplot of the first two components of the data shown in Fig. 3.

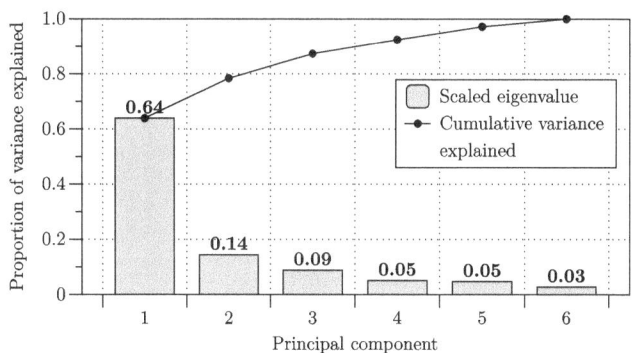

Fig. 5. Scree plot of the PCA whose first 2 components are shown in Fig. 4.

The comparison to the data points from the ideal group C_0 is interesting. A first observation is that C_1 is closer to the ideal course than C_2, but nevertheless, C_0 has a wider spread along the first component. Second, the most striking difference between C_0 and C_1 is along the WithComputer axis, indicating a tendency to consider that some part of CS teaching, contrary to what is being done now, should be done with paper/pencil. Roughly spoken, the ideal course focuses on the concept and notions like the C_1 optional course does now, but with a bigger emphasis on the know-how, and with a portion of it given outside the computer rooms.

Regardless of the distance between their ideal representation and the courses they are actually giving, respondents have a positive feeling towards CS teaching. Only one out of 37 respondents indicated being not satisfied with it, all others being either somewhat satisfied, satisfied, or very satisfied. 15 respondents (41 %) would like to teach CS more and only one would like to teach CS less (the others [21 people, 57 %] are satisfied with the current situation). Moreover, 92 % (33 out of 36) find it somewhat opportune, opportune, or very opportune for CS (as a science) to be taught to all students mandatorily.

Question 4. We wanted to know on what topics CS teachers needed continuing education. Two cases were distinguished: 1. the need for continuing education today in the context of the CS courses currently given (C_1 and C_2); and 2. the need for supplementary education that would arise if C_0 existed as a fundamental CS course for everyone (hereafter and in the legends referred to as "CS for all"). In the former case (today's situation), almost 90 % said they would need continuing education. In the latter, 79 % of those who said they would be interested to teach CS for all indicated they were likely to participate in a supplementary education program. Of those, nearly half (9 out of 21) said that they were even willing to participate in a program requiring about 300 hours of work (10 ECTS credits).

We are attached to a university for teacher education, and traditionally, we are not supposed to educate in matters related to the core discipline the future teachers will teach—only in matters of pedagogy and didactics. However, in certain fields, the need for courses with contents from the disciplines themselves is tangible. Thus, we first asked respondents to indicate the proportion of didactic aspects vs. aspects from the CS discipline they wanted to appear in the continuing education. The results, shown in Fig. 6, show that both now and in the hypothetical case of a future CS for all course, the continuing education courses offered to them should clearly not only consist of pedagogical aspects, but should review aspects from the fundamental CS discipline, too—and that even in a proportion slightly in excess of 50 %. This is interesting in two ways—fundamental scientific aspects are needed while pedagogical aspects are not dismissed as secondary or unimportant either.

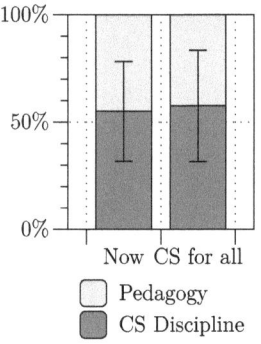

Fig. 6. Wanted breakup of continuing education related to CS.

Finally, we were interested in a list of topics for this continuing education that respondents would find most relevant and useful. Both for the current situation and in the case of a CS for all course, we asked them to grade topics as either unimportant, rather unimportant, rather important, and important. The number of respondents finding each topic at least rather important is shown in Fig. 7, with the topics being ordered according to the average awarded importance. The topics themselves are categorized in three groups, represented by different colors:

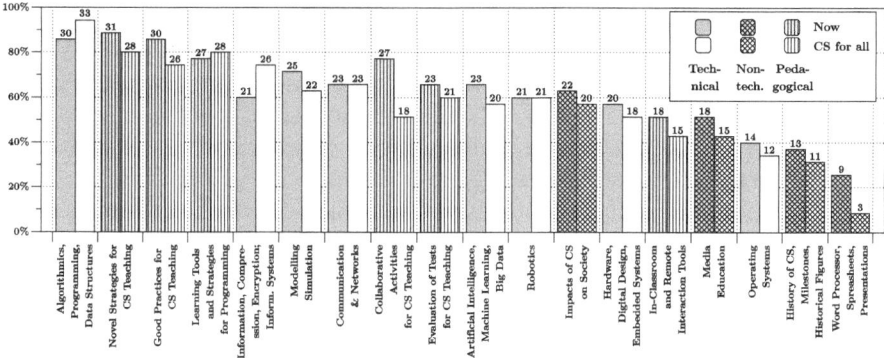

Fig. 7. Proportion of respondents who find the list of shown topics important in their continuing education as CS teachers.

first, fundamental topics from core CS; then, less technical topics related to the interaction of CS with society and media; and finally, topics linked to pedagogy and didactics.

A first observation is that the importance of topics is quite stable in the two distinguished cases. When considering the difference in awarded importance between the "now" and "CS for all" cases, we note a small, but statistically significant increase in importance for the group of core CS topics at the expense of the other two groups ($H(1) = 3.07$, $p = 0.079$). These results still give us a strong basis for the planning of continuing education courses today whose structure will still be relevant if and when CS for all is introduced.

We see that Algorithmics, Programming, Data Structures is the theme deemed most important, followed by three pedagogically oriented themes. Among the more technical themes, we can also observe that more importance is awarded to the fundamental themes (like programming, communication, representation of information) than to the more applied themes (like machine learning, robotics, operating systems). It remains an open question to know whether this is due to the fact that the respondents feel that the applied themes are less important in the context of their teaching, or that they feel they are more easily able to catch up on their own on such applied themes.

Then, whether we take the first 2, 6, 8, or 10 topics according to their awarded importance, we exactly have half of them belonging to the CS discipline and half of them treating pedagogical aspects, qualitatively reiterating the results from Fig. 7: the proposed continuing education should definitely not exclusively focus on pedagogical aspects to be of interest to the respondents.

5   Conclusion

We described the current state of CS teaching in Swiss high schools as well as some of the intricacies that led to it. Starting from that, we exposed our research questions, which we investigated with a survey sent to CS teachers from the French-speaking part of Switzerland.

The major results from our survey showed that most teachers currently do not have a primary education in CS, although the proportion in increasing. Most of them, however, had some form of complementary education in CS-related topics.

Regardless of their background, the respondents' view of what CS is does not differ fundamentally along the dimensions we explored, even if those with no CS education were less likely to have strong opinions on the nature of the field.

The representation of the ideal CS course, which gathers a strong agreement among respondents, differs from the two course types that are currently offered. The current offerings consist of a cantonal course, which is deemed as too focused on the usage of some software tools and not focused enough on concepts and notions, and of an optional course, which is closer to the ideal representation of the ideal CS course—one major difference being that the ideal course should include a more important part of pencil/paper activities and happen less often in front of a computer.

Finally, respondents indicate strong need for continuing education with a balanced proportion of both pedagogical topics and topics linked to the fundamental aspects of CS.

Our survey was only run on the French-speaking part of the country: it would be very interesting to extend this study to the German-speaking part of Switzerland, too, and to look into the causes for potential significant differences.

References

1. Bishop, C.M.: Pattern Recognition and Machine Learning. Springer, Heidelberg (2006)
2. Conférence suisse des directeurs cantonaux de l'instruction publique (CDIP): plan d'études cadre pour les écoles de maturité (1994)
3. Conférence suisse des directeurs cantonaux de l'instruction publique (CDIP): informatique au gymnase: remise d'un mandat pour l'établissement d'un rapport (2013)
4. Conseil fédéral, Conférence suisse des directeurs cantonaux de l'instruction publique (CDIP): ordonnance du Conseil fédéral/règlement de la CDIP sur la reconnaissance des certificats de maturité gymnasiale (RRM) (1995)
5. État de Fribourg: Plan des études gymnasiales, Domaine des branchescantonales, Informatique (2015). http://www.fr.ch/s2/files/pdf77/fr_maturite_gymnasiale_informatique.pdf
6. Mogey, N.: So you want to use a Likert scale. Learning Technology Dissemination Initiative 25 (1999). http://www.icbl.hw.ac.uk/ltdi/cookbook/info_likert_scale/
7. État de Vaud: Plan d'études de l'école de maturité (2015). http://www.vd.ch/fileadmin/user_upload/organisation/dfj/dgep/dgvd/fichiers_pdf/PET_EM.pdf

Work in Progress

IT2School – Development of Teaching Materials for CS Through Design Thinking

Ira Diethelm$^{(\boxtimes)}$ and Melanie Schaumburg

Computer Science Education, Carl von Ossietzky University,
26111 Oldenburg, Germany
{ira.diethelm,melanie.schaumburg}@uni-oldenburg.de

Abstract. Design Thinking is known as a process to create ideas and new applications. We were curious about the results when it was applied to the challenge of developing teaching materials for a non-profit project to support CS and IT (information technology) in schools from late primary to secondary schools. Therefore it had to take into account many different teachers and other actors. Our aim was to create usable and meaningful material that most teachers would like to teach with in their classes, regardless of their background or experience or knowledge of CS. Therefore it had to be interesting and powerful and at the same time easy to use and understand. And it also had to be motivating and inspiring for students aged 10 to 16. In this paper we report on our process and first insights. We also present some categories of teaching materials that came up during this process and personas of teachers. These may be helpful in other similar projects as well.

Keywords: Design thinking · CS teaching materials · Personas

1 Projects to Support CS

Although Computers and information technology in general are already part of our everyday life, CS is in many countries not a compulsory subject at schools. Therefore many projects are created to support CS or Informatics or Computational Thinking or programming at schools, like 'Hour of Code' [4], 'code.org' or the 'CoderDojo' [2], the 'Beaver Contest', 'CS Unplugged' [3] and lately the 'Micro:Bit' [1] just to name a few that are known internationally. But there are also very many national or regional activities that focus on supporting CS as a subject at schools or want to engage children and youths in creating their own pieces of information technology to play with and do something meaningful with. These are often initiated by private persons and also sometimes run in cooperation with local companies or local communities of software developers.

Most of these projects focus on one topic or method, like 'programming' or 'unplugged' or 'learning to program' or 'creating projects with a certain micro controller'. Sometimes they focus on environmental topics like geo-data, energy or light. Therefore, they tend to give a focused but maybe narrow view on CS at school. And most of these projects address students directly or are run outside of school and do not address whole schools and regular classes in the first place.

© Springer International Publishing AG 2016
A. Brodnik and F. Tort (Eds.): ISSEP 2016, LNCS 9973, pp. 193–198, 2016.
DOI: 10.1007/978-3-319-46747-4_16

2   Challenge

The German association 'Wissensfabrik' is a widespread and stable network of over 120 different big and small companies with long tradition of cooperation between companies and schools for supporting STEM and economical thinking inside the regular school schedule. As one of their scientific partners we were faced with the task to design and develop teaching material for the project 'IT2School', meeting each of the following requirements:

- to raise the motivation of children and youths for CS and IT,
- to give a wide insight into the variety of CS and IT in our everyday life,
- to lead to an understanding of some important principles of CS and to a stronger self-efficacy of students and teachers regarding CS and IT,
- to fit most grades at secondary schools, maybe even late primary schools,
- to be taught by primary and secondary school teachers who do not have a CS background and yet be interesting for those who have,
- to be taught inside other subjects like e.g. physics or art due to the fact that CS is often not a compulsory subject, and
- to be used in cooperation with local companies and their staff who also may not have a CS background or may have different notions of what might be important to teach in schools.

This challenge appeared quite unsolvable with other common methods inside a time frame of a year and a half of development time. So we tried Design Thinking to approach a most suitable solution.

3   Design Thinking

Design Thinking is a solution oriented process and also a toolbox with different methods. Design Thinking helps people to think like designers: Before creating a product, you should first try to understand the problem or issue by observing and talking with customers. Usually models or prototypes are created to be validated with the target group. The creative part of design is to create ideas to solve a problem, not to implement the solution. And one important point in Design Thinking is not to fall in love with the ideas, to allow to fail early and often, see [7].

Design thinking is an intentional and iterative process in order to achieve new, relevant solutions with positive impact. Design Thinking gives faith in creative abilities and a process for transforming difficult challenges into opportunities for design. And it is a solution oriented process - the mindset of Design Thinkers is that a solution can be found for any problem.

The design process is also a structured approach to generating and evolving ideas. It has six phases shown in Fig. 1 that help navigate the Design Thinking team from identifying a design challenge to finding and building a solution. After you have iterated your prototypes and when you think you and your target group

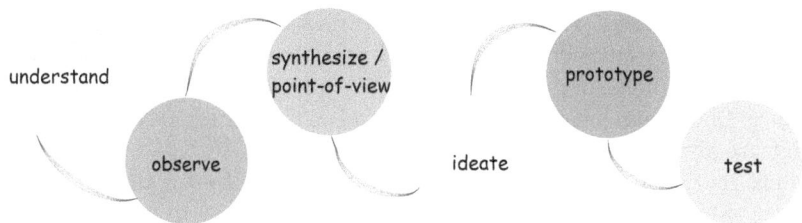

Fig. 1. Design Thinking process

are satisfied you can step into the 'implementation phase' - which is not a Design Thinking phase, but the next step to bring your proposed solution into real life.

It is a deeply human approach that relies on the ability to be intuitive, to interpret what you observe and to develop ideas that are emotionally meaningful to those you are designing for. The process is iterative and if necessary you can start at each phase again or go back to each phase. It is also collaborative, optimistic and experimental. That means, it begins from deep empathy and an understanding of needs and motivations of people, their goals and pains. To try to solve the problem in a team and benefits greatly from the views of multiple perspectives, and others' creativity. Design Thinking is the fundamental belief that we all can create change - no matter how big a problem, how little time or how small a budget. Irrespective of any existing constraints designing can be an enjoyable process. It gives permission to fail and to learn from mistakes as it is an iterative process, see [7].

4   Personas

One of the most important points in the design thinking process is the definition of 'points of view' or so called personas. Personas are detailed but fictional descriptions of certain different types of users or customers. In the first two phases of the design thinking process the needs of the target group(s) are in focus to come very close to good personas. For our task the teachers are the most crucial part. They are the ones who decide what to teach and when in what class. And they also decide whether they cooperate with a company or not. So first we put our focus on them.

With our experience and considering the research on teacher education in general and on CS teachers in particular, and also research on teacher personalities, see [10] in several iterations of step 1 and 2 we came to the following points of view:

Sabrina: She is extroverted and wants to save the world. Her intention in teaching CS is to engage all students and also to show her colleagues what is possible. She is heading for being a teacher educator or principal. She needs a project that results also in reading her name in the local newspapers and meet interesting people to network with from outside of school.

Susan: She is a very reflected and well-educated and a slightly introverted teacher who is emotionally connected with teaching and will always be a teacher. Her main aim is to prepare all students as well as possible for the future. She needs meaningful content and enlightening ideas for her lessons that are always in progress.

Robert: He did not plan to be a teacher, he once missed a chance to become a professional football player. But now he has a family to care for and many other things on his mind. He cannot afford to spend much time on preparation or rethinking lesson plans. He needs something that works more or less out of the box and ensures that the students are occupied for some hours so that he can think of something else.

Tim: He is a self-conscious tinkerer who teaches physics and technology. He is also interested in computers but mostly in hardware. He is open-minded and believes that if some students of his class gets the ideas and the motivation the others will follow. He needs only a few hints and some hardware items as an example. Then he will try it in his class and adjust the intended curriculum to his needs spontaneously.

Irene: She is a very experienced teacher who really wants the students to learn a lot and therefore she is a bit afraid of failing with new content inside her classroom. She is quite open but only likely to try something if it is written down in detail what to do when and why. She will try it only if she is convinced that it will work for all students in her class.

Herbert: He is a very experienced teacher who likes to keep control over the class and likes a clear outline of what is done in which lesson. He likes tradition and is not likely to change his curriculum that he has taught for many years now. He also has little interest in new or fancy technology.

5   Ideas and Prototype

With these personas in mind we went into the 'ideate' phase of the Design Thinking process. In this phase all ideas are allowed and welcome: the safe bets, the crazy ideas and also delighting ones. It is not allowed to argue against any idea you or your team brings up in this phase. Every idea is written on a post-it note and put on a wall. We collected many ideas from existing projects and also came up with a few new ones for CS at school that would meet these needs and allow an authentic teaching style.

From our experience in teacher education and also from [9] we derived two dimensions on how teachers and especially our personas would judge teaching material at first sight:

Look and Feel: What does the material look like? What is the nature of the materials it analogous or paper-based or with software or with hardware pieces or are other places or persons from outside of school part of it? (right)

Student Activity: How do the students interact with it? What is the teacher's role - is it to teach or to guide? Are the classroom activities focusing on understanding, detecting principles or recognizing them or are the students exploring, trying out? Or do students create something on their own and share it with other people, parents, friends or relatives? (up)

In a next step we sorted every post-it note into this matrix and tried to find at least one idea for every square in the matrix, see Fig. 2.

share				
create				
explore				
understand				
activities / look and feel	analogous / unplugged	with software	with hardware	with external partners or places

Fig. 2. Matrix for the ideation phase based on personas

Then, we tried to validate the teachers' needs and to anticipate what each of the personas would understand and like or dislike about each idea. In this phase many other people, mostly teachers, who came by our office were asked what they thought how each persona would deal with the ideas and how the requirements are met. Only those ideas resulted as one of the basic modules that would fit all personas. It was not necessary that they would all use the same module in the same way. As a first result we found five basic modules:

1. Communication from blinking to encryption: This modules starts at coding information via blinking, see also [5], and the historical dimension of transporting information over large distances by Morse code, also building a string telephone and submitting pictures are part of the experience [3,6] and it ends with easy Caesar's encryption.
2. Understanding the internet: How does such a big world fit into such a tiny box? This question guides a setting to explain how the internet works, with paper and other craft materials only, but also using IP-addresses.
3. Codes in a supermarket: How does the checkout know the price? What happens if we manipulate barcodes and how are these used to run a supermarket are discovered in this module. Here, students and teacher should visit a local supermarket and get a view behind the scenes. Afterwards students think about useful innovations using QR-Codes in their school, e.g. for a game.
4. How to program, using Scratch, also must be part of such a setting and is also usable for all kinds of personas.
5. My very special input: What happens if we connect bananas or play-doh to the computer and use them as a keyboard is discovered in this module using Makey Makeys [8] or similar controllers.

For these basic modules we created prototypes to test with teachers, students and members of cooperating companies. This sample consists of persons from the Wissensfabrik's network and from our teacher trainings. This process is constantly taking place and we are collecting feedback from each of these groups and refine our ideas and prototypes.

6   Test and Next Iteration

We continuously iterate the Design Thinking phases. While collecting feedback we learn more and more about our target groups. Resulting from feedback that we collected in the first iteration we noticed another need for two more basic modules: One was to add a module for the methodology of Design Thinking itself. This is not suitable for every teacher but many of our (open-minded) teachers we worked with were interested in how we created the modules.

The second need was to add something that was a bit more challenging and had to do with smartphones. So we created a module to programm one's own app with AppInventor to control a light via bluetooth. This is a module for teachers like Sabrina who are already quite confident and search for something special. All module descriptions for present basic modules are already available for free at the project's webpage www.it2school.de in German.

The project is (and maybe always will be) work in progress. More modules are planned: some for smaller children and some more challenging ones that go deeper into CS regarding security issues or exploring big amounts of data. We are now also creating personas for the employees of companies who co-operate with these teachers to create training sessions for them.

References

1. BBC - The Micro Bit - home. https://www.microbit.co.uk/. Accessed 09 June 2016
2. CoderDojo. https://coderdojo.com/about/. Accessed 09 June 2016
3. Computer Science Unplugged. http://csunplugged.org/. Accessed 09 June 2016
4. Hour of Code. http://hourofcode.com/. Accessed 09 June 2016
5. Curzon, P.: Computational Thinking: Searching to Speak — Teaching London Computing. http://teachinglondoncomputing.org/free-workshops/computational-thinking-searching-to-speak/
6. Hunkin, T.: The Secret Life Of The Fax Machine. http://www.secretlifeofmachines.com/secret_life_of_the_fax_machine.shtml. Accessed 09 June 2016
7. Ideo, L.L.C.: Design Thinking for Educators, 2nd edn. Riverdale, Riverdale (2012)
8. JoyLabz. Makey makey. http://makeymakey.com/. Accessed 09 June 2016
9. Stoffers, A.-M., Diethelm, I.: Teacher profiles for planning informatics lessons. In: Gülbahar, Y., Karataş, E. (eds.) ISSEP 2014. LNCS, vol. 8730, pp. 150–160. Springer, Heidelberg (2014)
10. Terhart, E. (ed.): Handbuch der Forschung zum Lehrerberuf, 2nd edn. Waxmann, Münster (2014)

"Why Can't I Learn Programming?" The Learning and Teaching Environment of Programming

Zsuzsanna Szalayné Tahy[1(✉)] and Zoltán Czirkos[2]

[1] Faculty of Informatics, Eötvös Loránd University, Budapest, Hungary
sztzs@caesar.elte.hu
[2] Department of Electron Devices,
Budapest University of Technology and Economics, Budapest, Hungary
czirkos@eet.bme.hu
http://inf.elte.hu, http://vik.bme.hu

Abstract. This article focuses on teaching a textual programming language as the first programming language (allowing for previously studied visual programming languages). The teaching process is placed into a real educational environment in connection with the national curriculum, social expectations and students reactions. In order to write a program, several abilities and pieces of knowledge are required. There are tools and syllabuses for teaching these skills but the result mainly depends on the personality of the students and teachers. We use the term "Learning Activity Unit" to describe the teaching–learning process of programming and detecting gaps in every day practice. This very simple model is practical for teachers to detect problems. In the global view of teaching programming, the implementation of the curriculum could be analysed.

Keywords: Computational thinking · Curriculum design · Programming · Teaching-learning process · Learning Activity Unit

1 Introduction

Twenty years in practice gives many impressions in teaching informatics. Focusing on textual programming, the main concepts are almost the same, but the tools have changed. The choice of language has changed from Pascal to C-based languages and nowadays the object oriented programming concept is preferred. Several methods have been tested to improve the effectiveness of teaching. At the beginning it was 10–20% of the students who had to learn programming, nowadays this number should approach 100 % [3]. The extension of teaching programming skills raises new question: Is everybody able to learn programming?

Teachers have some practice based on other subjects to use special methods in several cases. There are studies on how to teach some skills in a selected, already-known group, or well-known prior knowledge/abilities [1]. However, the digital age is too young to know all the answers.

The Learning Activity Unit is a diagnostic tool used for analysing expectations and individual achievement of knowledge and skills required to develop

© Springer International Publishing AG 2016
A. Brodnik and F. Tort (Eds.): ISSEP 2016, LNCS 9973, pp. 199–204, 2016.
DOI: 10.1007/978-3-319-46747-4_17

computational thinking [4]. This article presents some sample cases to demonstrate how this tool can be used in the learning environment analysis. Curricula, syllabuses and lesson plans can be analysed by the Learning Activity Unit concerning aims, methods and timing. This analysis gives a global view of the learning process, guiding the long-term work. The presented collection is far from being complete but the examples describe real problems or contra-productive practices and their recommend solutions.

2   Programming – The Learning Activity Unit Framework

In this section we present the Learning Activity Unit (LAU), a framework whose main objective is to understand the learning process of programming, and to diagnose learning problems. The Learning Activity Unit encompasses the complete learning cycle, from the starting point, when the students first meet a concept of programming, to the point when they are able to use it on their own. The phases defined by the model are as follows:

1. **Initial learning**: As the first step, programs, solutions of programming tasks are written by following the instructions.
   - (A) **Active** learning: students are motivated to learn, they read sources, listen to lectures. They follow the logic of problem solving.
   - (M) **Moderated** learning: students make notes; they observe what the teacher does and try to do the same.
   - (P) **Passive** learning: students are outside observers, they scroll the readings.
2. **Trying** phase: students (or the teacher) explore how the new knowledge or skill could be used.
3. **Experimenting** phase: students have to change the written program, or they have to solve some very similar tasks with the help of teacher.
4. **Pause**: some learned details could be forgotten.
5. **Using**:
   - a. **Repeating** means repeating the learned things in phase 1.
   - b. **Modifying**: students are able to learn creativity, but only at the original level presented.
   - c. **Creating** new programs could give the sense of success.
6. Back to phase 4 or phase 1. It implies a lifelong learning process; however, the process could be broken in every phase.

2.1   The Using of the Learning Activity Unit

The LAU is not homogeneous and not absolute. It is a practical tool to focus on the main points of the learning and teaching process of programming. We apply LAU in a similar way to how programmers work with functions and threads. Every LAU allocates a part of capacity in the human' mind. There are LAUs running parallel, calling each other and embedded into each other. The LAU Model – with the relationships between LAUs – describes a complex learning process.

There are many cases in the preparatory period of learning textual programming, where we can observe this activity model with the return to phase 1, for example learning a visual language or learning how applications work. If a curriculum is well-structured, these studies develop many important skills and always gives new knowledge to the student. On the other side, there are many courses, programming actions what provide only a foretaste of the knowledge, a short insight, but no more. There are courses focusing on phases 1–3. These courses seem to be very successful, but the outcome is useless.

For illustration only, a humanities educated parent wrote: "My 12-year old daughter is very good in logical tasks. She does programs in Scratch but follows always the trodden path... She attends courses but it seems there is no novelty for her." However, we do not explore those courses, but this opinion shows two problems: (1) This girl prefers 5a **Repeating** activity and courses do not motivate her to choose 5c **Creating**. (2) These courses are only good for wakening up interest for programming, but cannot improve or develop the knowledge to a usable level. Even if courses could develop skills, they are not in connection, they are not structured therefore every course starts from the same basic level.

The LAU seems to be simple enough to use in every day practice, nevertheless, it fits to Bloom's renewed taxonomy and takes into consideration the effectiveness of learning methods as well as the forgetting rates and benefits of linear and spiral curriculum design.

3   Programming Curriculum in Hungary

According to the Hungarian National Curriculum[1], students are required to learn programming. The advanced level secondary school final exam in Informatics includes a task testing the algorithmic and programming skills. The history of informatics education is similar to the Polish system, described in 2015 at the ISSEP conference [5]. The Hungarian National Curriculum was accepted in 2008 but it was renewed in 2012, expanding knowledge expectation compared to the former version. Skill expectations are similar to the new Polish Curriculum but supplemented with topics of hardware and network knowledge (e.g. ISO OSI Model). Unfortunately, the Core Curriculum[2] – prescribed by government – cut the number of lessons to the third compared to the National Curriculum. However, the Hungarian IT sector, the representatives of universities and civil groups (e.g. parents) demand an increase number of lessons of Informatics.

The elimination of informatics lessons has two consequences. On the side of public education, skills and knowledge is to be learnt in only one third of the required time. Although this seems to be nonsense, it is written in the certified syllabus[3]. Analysing the syllabus, timing limits teaching to the list of concepts.

---

[1] Nemzeti alaptanterv (National Curriculum), http://www.kozlonyok.hu/nkonline/ MKPDF/hiteles/MK12066.pdf, Magyar Közlöny vol 66 (2012) (in Hungarian).

[2] KT 9-12G (Core Curriculum of informatics for grade 9–12) http://kerettanterv.ofi. hu/03_melleklet_9-12/3.2.16_informat_9-12.doc (2012) (in Hungarian).

[3] Informatika 10. tanmenet (Syllabus for grade 10) http://ofi.hu/sites/default/files/ attachments/nt_17173_informatika_10.docx (2016) (in Hungarian).

Students hear (or do not hear) the concepts but there is no time to practice them. Described in LAU terms, this is only part of the **Initial learning** (1), because it is based on informal learning, too. The **Trying** (2) is homework, the **Experimenting** (3) and sometimes the **Using** (5) would be part of other subjects. In many cases, the **Pause** phase is too long, or there is too much time between the next **Initial learning** and previous **Using** phase.

On the other side, many companies from the IT sector try to supplement the programming lessons, to fill in the gaps of public education. But this effort cannot reach the goal because they are not able to ensure long-term development. A 10-hours crash course, or a 30-hours weekend-only courses, maybe a one-week-long camp or a half-year-long course in learning programming gives "a sneak peek". It looks very good, but these courses are not connected to each other, therefore long-term effectiveness is uncertain. In the view of LAU: Phases 1–3 are prepared but phase 4 is too long, phase 5 may never come. The return (loop back to phase 1) will result in random development or backwardness. Moreover, this practice is very dangerous at the point when governmental education management envisions the teaching of informatics as activities of summer camps.

4   Introduction to Textual Programming

4.1   Expected Skills and Hidden Gaps

In order to code a program, one needs almost a dozen skills. Studies of teaching programming explore the role of these skills; describe methods of developing one or a group of skills [6]. Modern educational systems offer opportunities to improve these skills before learning textual programming. Skills and knowledge are mentioned in different ways according to the focus of research. The following skills were found useful to learn before text-based programming: (1) typing, (2) mother tongue based comprehension, (3) basic reading and writing in English, (4) practice in multi window software using, (5) abstraction, (6) logical decision, (7) recognizing and defining data types (boolean, character, integer, real, string), (8) recognition and defining data structures (array, 1D, 2D, 3D, record, graph), (9) object modelling, (10) algorithmic thinking (sequence, alternation, loop) in real word situations, (11) understanding and using functions of applications (e.g. text editor, spreadsheet, animation designer), (12) system (process) modelling.

The list, of course, may not be complete, but the more important aspect is the knowledge level of skills. When one writes a for-loop, six skills are activated from the above mentioned ones, and every, missing skill is a gap. As it is described in [2], "small steps" are very important in effective teaching. The authors of this article analysed books and described a tool for detecting gaps in textbooks, but practically there were no textbooks without big gaps. It seems that text-based programming is too complex, and the success of teaching text-based programming depends on how many items are known before using them.

This problem can be observed at the Basics of Programming 1 course[4] of Engineering Information Technology at BME. We conducted an experiment in 2014, asking 3–10 questions from 525 students every week about the topic of the lecture. For example, after the second lecture, 225 students answered this question about variables:

Have you ever heard about variables before this lecture?
1. I haven't ever heard about them, this is new for me.
2. I have heard about them, but I've never tried them.
3. I have seen, I've tried in some cases
4. I have used this knowledge, I am experienced.

The average result of the first test written by students who used variables before the course (i.e. who had chosen answers 3. or 4. in the questionnaire) was 72 %. On the other hand, those who had chosen answers 1. or 2. only scored 46 % in the test. The lack of prior knowledge caused difficulties in their learning progress.

In September 2015 we asked the 565 new-coming students to fill in a 26-item questionnaire about the input skills and knowledge and some question about the learning habits. There were 346 students who answered 77 % of the questions on average. We correlated the answers with the test results, created a table of 26 rows (the questions) and 13 columns (the test results) with values between 0.45 and −0.13. Selecting the highest three values from every column (for the tests of different topics during the semester) we get the highest ranking, most correlating questions. This way we found that the most relevant prior skills are:

1. Knowing data structures (12) – What kind of data structures have you used?
2. Knowing Code::Blocks (11) – Have you used Code::Blocks before?
3. Programming (6) – How many points did (could) you get in the secondary school final exam's programming task?
4. Maths knowledge (6) – What mark did you get in the Maths exam?
5. Algorithms (2) – What kind of algorithms have you learnt?
6. Physics knowledge (1) – How would you mark your physics knowledge?

Physics was in connection with the homework, the students' own programmed game. We asked about music, spreadsheet, databases, Nassi-Shneiderman chart, languages, grammar and other topics as well. We can say that the most correlated skills are the most relevant in the course.

We would like to extend our research to the Eötvös Loránd University, Faculty of Informatics. There are also almost 600 students but the courses are in Maths science while BME is the centre of engineering education. The gathered information would be very important to determine programming expectations in secondary schools.

Even though we still have to clear the details, we can say at this moment that the compensating the lack of prior knowledge needs more time. It involves the multiple usage of the LAU, and the preparation phase should be longer for successful teaching. The formal courses of Introduction to Computer Science, Basics of Programming or Introduction to <any text-based> Programming Language starts with a huge gap for real beginners.

---

[4] Z. Czirkos, G. Nagy: INFOC Portal for course https://infoc.eet.bme.hu/.

4.2 De-gap Before Start

Teaching programming must be preceded by a long preparatory period, when students learn particular competences. This period starts at the beginning of education and the effectiveness depends on the awareness of educators. Many people – students, parents, teachers and experts among them – say, the preparatory period is not for programming. However, as cooking starts with shopping the ingredients, learning textual programming starts with learning the necessary skills. Shopping is much more effective if you know what you want to cook. By the analogy, it would be very useful if primary school teachers would be able to write codes. Not to actually teach programming, but to understand how they should teach the basics.

5 Summary

Textual programming is based on several abilities and skills. The successful learning programming requires the creative usage of basics, therefore text-based programming should be preceded by a designed preparatory period. We described a Learning Activity Unit model based on main concepts of pedagogy to characterize the skill level of computational thinking. Having applied it for analysing Hungarian curricula and courses, we detected problems of effectiveness: only **Initial learning** is planned but sometimes it is also compacted. Our further research will focus on teaching practice of programming in classroom.

References

1. Heintz, F., Mannila, L., Nygårds, K., Parnes, P., Regnell, B.: Computing at school in Sweden - experiences from introducing computer science within existing subjects. In: Brodnik, A., Vahrenhold, J. (eds.) ISSEP 2015. LNCS, vol. 9378, pp. 118–130. Springer, Heidelberg (2015)
2. Hofoku, Y., Cho, S., Nishida, T., Kanemune, S.: Why is programming difficult? - Proposal for learning programming in "Small Steps" and a prototype tool for detecting "gaps". In: ISSEP 2013, pp. 13–14. Universitätsverlag Potsdam (2013)
3. Informatics Europe & ACM Europe Working Group on Informatics Education: Informatics education: Europe cannot afford to miss the boat (2013)
4. Lee, I., et al.: Computational Thinking Resources (2011). https://hcsta.acm.org/Curriculum/sub/CompThinking.html. Accessed 21 July 2016
5. Syslo, M.M., Kwiatkowska, A.B.: Introducing a new computer science curriculum for all school levels in Poland. In: Brodnik, A., Vahrenhold, J. (eds.) ISSEP 2015. LNCS, vol. 9378, pp. 141–154. Springer, Heidelberg (2015)
6. Szlávi, P., Zsakó, L.: Methods of teaching programming 1(2). In: Teaching Mathematics and Computer Science, 1.02, pp. 247–258. University of Debrecen, Hungary (2003)

Author Index